The Effects of Credit Policies on U.S. Agriculture

T0272902

AEI STUDIES IN AGRICULTURAL POLICY

AGRICULTURAL POLICY REFORM IN THE UNITED STATES
Daniel A. Sumner, ed.

AGRICULTURAL TRADE POLICY: LETTING MARKETS WORK
Daniel A. Sumner

ASSESSING THE ENVIRONMENTAL IMPACT OF FARM POLICIES
Walter N. Thurman

CHOICE AND EFFICIENCY IN FOOD SAFETY POLICY
John M. Antle

THE ECONOMICS OF CROP INSURANCE AND DISASTER AID
Barry K. Goodwin and Vincent H. Smith

THE EFFECTS OF CREDIT POLICIES ON U.S. AGRICULTURE
Peter J. Barry

MAKING SCIENCE PAY: THE ECONOMICS OF AGRICULTURAL
R&D POLICY
Julian M. Alston and Philip G. Pardey

REFORMING AGRICULTURAL COMMODITY POLICY
Brian D. Wright and Bruce L. Gardner

The Effects of Credit Policies on U.S. Agriculture

Peter J. Barry

The AEI Press

Publisher for the American Enterprise Institute

WASHINGTON, D.C.

1995

Library of Congress Cataloging-in-Publication Data

Barry, Peter J.
    The effects of credit policies on U.S. agriculture / Peter J.
Barry
        p.   cm. — (AEI studies in agricultural policy)
    Includes bibliographical references and index.
    ISBN 978-0-8447-3905-2 (pbk)
    1. Agricultural credit—Government policy—United
States. 2. Agriculture and state—United States. 3. Ag-
riculture—United States—Finance.   I. Title.   II. Series.
HD1440.U6B37   1995
338.1'0973—dc20   95-17826
                                                              CIP

1 3 5 7 9 10 8 6 4 2

The AEI Press
Publisher for the American Enterprise Institute
1150 17th Street, N.W., Washington, D.C. 20036

Contents

LIST OF FIGURES

Foreword

The Effects of Credit Policies on U.S. Agriculture by Peter J. Barry is one of eight in a series devoted to agricultural policy reform published by the American Enterprise Institute. AEI has a long tradition of contributing to the effort to understand and improve agricultural policy. AEI published books of essays before the 1977, 1981, and 1985 farm bills.

Agricultural policy has increasingly become part of the general policy debate. Whether the topic is trade, deregulation, or budget deficits, the same forces that affect other government programs are shaping farm policy discussions. It is fitting then for the AEI Studies in Agricultural Policy to deal with these issues with the same tools and approaches applied to other economic and social topics.

Periodic farm bills (along with budget acts) remain the principal vehicles for policy changes related to agriculture, food, and other rural issues. The 1990 farm legislation expires in 1995, and in recognition of the opportunity presented by the national debate surrounding the 1995 farm bill, the American Enterprise Institute has launched a major research project. The new farm bill will allow policy makers to bring agriculture more in line with market realities. The AEI studies were intended to capitalize on that important opportunity.

The AEI project includes studies on eight related topics prepared by recognized policy experts. Each study investigates the public rationale for government's role with respect to several agricultural issues. The authors have developed evidence on the effects of recent policies and analyzed alternatives. Most research was carried out in 1994, and draft reports were discussed at a policy research workshop held in Washington, D.C., November 3–4, 1994.

The individual topics include investigation of:

- the rationale for and consequences of farm programs in general
- specific reforms of current farm programs appropriate for 1995, including analysis of individual programs for grains, milk, cotton, and sugar, among others
- agricultural trade policy for commodities in the context of recent multilateral trade agreements, with attention both to long-run goals of free trade and to intermediate steps
- crop insurance and disaster aid policy
- the government's role in conservation of natural resources and the environmental consequences of farm programs
- farm credit policy, including analysis of both subsidy and regulation
- food safety policy
- the role of public R & D policy for agriculture, what parts of the research portfolio should be subsidized, and how the payoff to publicly supported science can be improved through better policy

Peter Barry deals with the complex array of government programs and policies related to farm finance. The issues range from what to do about large budget costs or exposure of the Farmers Home Administration (which has now been made a part of the new Consolidated Farm Service Agency) to how regulations and federal backing affect the private institutions that form the Farm Credit System. Reforms in the 1980s made major changes in farm financial policy, and some of the remaining problems are the legacy of an earlier era. But significant issues remain, and Barry examines carefully where the government's role may be modified and how policy can make better use of private incentives. One goal of credit policy improvements is to avoid the kind of debacle that characterized farm credit in the 1980s, when public policy contributed to financial disarray at a large cost to tax payers and farmers.

Selected government policy may be helpful in allowing agriculture to become more efficient and effective. Unfortunately, most agricultural policy in the United States fails in that respect. In many ways, the policies of the past six decades have been counterproductive and counter to productivity. Now, in the final few years of the twentieth century, flaws in policies developed decades ago are finally becoming so obvious that farm policy observers and participants are willing to consider even eliminating many traditional subsidies and regulations. In the current context, another round of minor fixes is now seen as insufficient.

In 1995, Congress seems ready to ask tough questions about agricultural policy. How much reform is forthcoming, however, and which specific changes will be accomplished are not settled and depend on the information and analysis available to help guide the process. Understanding the consequences of alternative public policies is important. The AEI Studies in Agricultural Policy are designed to aid the process now and for the future by improving the knowledge base on which public policy is built.

CHRISTOPHER DEMUTH
American Enterprise Institute
for Public Policy Research

1
Introduction

Public credit programs and financial policies have long played a significant role in agricultural finance through government loan programs, government-sponsored enterprises, and regulations of depository institutions. The general rationale for public intervention is usually based on two issues (Bosworth, Carron, and Rhyne 1987). First, private credit markets cannot meet the social objectives and priorities affecting the allocation of resources and distribution of income. Second, perceived imperfections in private credit markets result in credit rationing, market failure, and other types of credit gaps. In recent years, considerable attention in the professional literature has focused on public intervention motivated by problems of asymmetric information. These problems constrain the effectiveness of private credit markets and thus hamper efficiency and welfare attainment (Gale 1987, 1990a, 1990b, 1991; Mankiw 1986; Smith and Stutzer 1989; Stiglitz and Weiss 1981). Public intervention is costly, however, as indicated by estimates of annual subsidies and by the substantial assistance provided to financially troubled farmers and their lenders during the 1980s.

This monograph discusses farm financial policy with an emphasis on public credit programs and their effects

1

on the farm sector. The discussion begins by reviewing the major characteristics of U.S. agriculture and of its suppliers of financial capital. Next, perspectives on credit issues and programs in agriculture along with public credit programs are established. New theories of public credit motivated by asymmetric information problems are considered next. Then, three types of government involvement in agricultural finance receive attention:

- government credit programs, mostly serving as a lender of last resort for farmers, through the Farmers Home Administration at the federal level and similar programs at the state level
- government-sponsored enterprises providing credit and financial services directly to agricultural borrowers in the case of the Farm Credit System (FCS) and providing a secondary market for farm real estate loans in the case of the Federal Agricultural Mortgage Corporation (Farmer Mac)
- regulations of the U.S. banking system

A concluding chapter considers the general implications for the financing of a diverse agricultural sector and its major sources of credit.

2
Credit Characteristics of U.S. Agriculture

Credit demands in U.S. agriculture are determined by a broad set of structural and demographic characteristics. These characteristics combine to indicate the sector's uses of financial capital, patterns of creditworthiness, and the roles of public policies in responding to credit market imperfections and in meeting social objectives for the agricultural sector. These credit characteristics are described as follows.

Capital Intensity and Low Liquidity

Agriculture is a capital-intensive industry in which investments in farmland, buildings, machinery, equipment, and breeding livestock dominate the asset structure of most types of farms. As shown in table 2–1, farm real estate has accounted for more than 70 percent of total assets in the sector since the early 1970s, reaching nearly 80 percent of total assets in the early 1980s, when prices per acre of farmland in the United States reached their peak. Inventories of livestock, machinery, crops, and other non–real estate farm assets constitute another 10 to 15 percent of the sector's total assets. The dominance of farm real estate together with the relatively small holdings of financial

3

TABLE 2-1
U.S. Farm Sector Balance Sheet, Including Operator Households, 1950–1990
(billions of dollars)

	1950 $	1950 %	1960 $	1960 %	1970 $	1970 %	1975 $	1975 %	1980 $	1980 %	1985 $	1985 %	1990 $	1990 %
Farm Assets														
Real estate	88.9	58.1	139.7	66.3	224.5	69.3	421.0	72.7	850.1	78.1	657.0	73.6	702.6	70.5
Livestock and poultry	17.1	11.2	15.6	7.4	23.7	7.3	29.4	5.1	60.6	5.6	46.3	5.2	69.1	6.9
Machinery and motor vehicles	14.1	9.2	22.2	10.5	34.4	10.6	63.1	10.9	86.9	8.0	88.3	10.0	91.7	9.2
Crops	7.1	4.6	6.7	3.2	8.4	2.6	21.2	3.7	31.9	2.9	22.9	2.6	22.4	2.2
Purchased inputs	NA		NA		NA		NA		NA		1.2	0.1	2.8	0.3
Household equipment and furnishing	9.6	6.3	8.7	4.1	10.0	3.1	14.2	2.5	19.4	1.8	27.8	3.1	46.3	4.6

	$	%	$	%	$	%	$	%	$	%	$	%	$	%
Financial assets	16.1	10.5	17.8	8.4	17.0	5.2	30.4	5.2	39.3	3.6	49.3	5.6	61.1	6.1
Total	152.9		210.7		324.0		579.2		1,088.2		892.8		996.0	
Farm debt														
Real estate	6.1	50.0	12.9	52.0	30.5	57.8	49.9	54.5	97.5	54.6	105.7	56.3	78.4	54.0
Non-real estate	6.1	50.0	12.0	48.0	22.5	42.2	41.6	45.5	81.2	45.4	82.2	43.7	66.7	46.0
Total	12.2		24.8		52.8		91.5		178.7		187.9		145.1	
Equity capital	140.7		185.9		271.3		487.7		909.5		704.9		850.9	
Debt/assets, %	8.0		11.8		16.3		15.8		16.4		21.2		14.6	
Debt/equity, %	8.7		13.3		19.5		18.8		19.6		26.9		17.1	

NOTE: Percentages may not add to 100 because of rounding. Percentages are percentages of totals; NA = not available.
SOURCE: National Financial Summary, 1990; *Economic Indicators of the Farm Sector*, U.S. Department of Agriculture, ECIF 10-1, November 1991.

TABLE 2–2
INDEBTED FARMS, LAND IN FARMS, AND OWNED OR RENTED
ACREAGE, BY REGION AND DIVISION, 1988

	Farms with Debt, %	Land in Farms (1,000 acres)	Acres Rented, %	Acres Owned, %
United States	52.8	852,184	45.4	54.6
Region				
Northeast	46.3	18,940	25.7	74.3
Midwest	60.1	328,788	48.7	51.3
South	45.9	251,263	43.7	56.3
West	54.2	253,193	46.6	53.4
Division				
New England	51.2	3,917	17.8	82.2
Mid-Atlantic	45.0	15,023	27.7	72.3
East north central	56.3	80,982	48.2	51.8
West north central	62.9	247,806	46.4	53.6
South Atlantic	47.5	44,720	31.5	68.5
East south central	42.1	39,642	28.3	71.7
West south central	47.6	166,901	50.7	49.3
Mountain	57.3	198,045	44.6	55.4
Pacific	51.6	55,149	54.0	46.0

SOURCE: Bureau of the Census, *Agricultural Economics and Land Ownership Survey* (1988), volume 3, 1987, Census of Agriculture, AG 87-R5-2, January 1991.

assets indicates the high capital intensity and low asset liquidity of the sector.

Capital Intensity and Low Debt-carrying Capacity

The aggregate debt-to-asset ratio of the farm sector increased steadily to reach 15 to 18 percent in the 1970s and then rose above 20 percent in the mid-1980s, as it reflected the decline in farm real estate values that characterized this period. Substantial reductions in aggregate farm debt and modest recoveries in values of farmland and other assets in the late 1980s brought this solvency measure back within 15 to 18 percent.

The farm sector debt-to-asset ratio of 15 to 18 percent appears low relative to many other economic sectors. Part of the difference reflects the use of current market values of farm real estate compared with original cost-adjusted book values for depreciable assets in other sectors. The lower range for the debt-to-asset ratio, however, is consistent with the heavy reliance in agriculture on a nondepreciable asset such as farmland in which much of its economic return occurs as capital gains or losses on real estate assets (Melichar 1979; Barry et al. 1995). As shown by several studies (Barry and Robison 1986; Ellinger and Barry 1987), the debt-carrying capacity of nondepreciable assets (for example, land) is considerably lower than that of depreciable assets. Thus, lower aggregate debt-to-asset ratios for the farm sector are logical to expect. Within the farm sector, however, debt-to-asset ratios may differ substantially according to a farm's tenure position, size, age of operation, and enterprise mixture.

Extensive Leasing of Farm Real Estate

In light of the dominance of farm real estate in the sector's total assets and its low debt-carrying capacity, the leasing of farmland by farm operators has become a widespread and commonly accepted method of financing that is especially effective for expanding farm size. In 1988, 45.4 percent of total farmland in the United States was operated by farmers under a rental arrangement with the landowner (table 2–2). The remaining acreage was farmed by an owner-operator. The incidence of leasing is highest in the Midwest and lowest in the Northeast.

The dominant form of rental arrangement in 1988 was a cash lease (65.3 percent) in which a fixed or flexible

TABLE 2–3
PERCENTAGE OF LEASED ACREAGE CONTROLLED UNDER VARIOUS
TYPES OF RENTAL ARRANGEMENTS FOR REGIONS
AND SELECTED STATES, 1988

	Acres (1,000)	Cash Lease	Share Leases	Cash/ Share Leases	Other Leases
United States	331,923	65.3	29.8	3.2	1.7
Region					
Northeast	4,747	86.3	5.6	0.8	7.4
Midwest	142,291	56.3	38.7	3.3	1.6
South	106,413	72.1	23.3	3.3	1.3
West	71,471	72.5	22.5	2.8	2.2
Division					
New England	693	73.4	18.5	1.6	6.5
Mid-Atlantic	4,054	88.5	3.4	0.7	7.5
East north central	38,906	50.5	45.7	2.5	4.9
West north central	110,385	58.3	36.3	3.6	1.7
South Atlantic	14,039	86.3	8.4	2.0	3.2
East south central	11,134	75.8	16.5	5.3	2.3
West south central	81,240	69.1	26.7	3.3	0.9
Mountain	49,603	73.6	20.4	3.1	2.9
Pacific	21,868	70.0	27.2	2.1	0.6
East and west north central states					
Ohio	5,605	61.9	35.8	0.6	1.7
Indiana	7,346	47.1	49.7	2.1	1.0
Illinois	18,286	33.7	61.3	3.6	1.4
Michigan	3,408	80.3	18.0	1.1	0.5
Wisconsin	4,262	89.5	6.7	1.5	2.2
Minnesota	10,869	74.3	20.6	1.5	3.6
Iowa	14,441	50.7	43.4	4.9	0.9
Missouri	8,606	48.6	42.7	4.8	3.9
North Dakota	20,442	64.3	31.9	0.6	3.3
South Dakota	12,962	65.7	28.1	5.6	0.5
Nebraska	18,387	68.5	27.7	3.3	0.4
Kansas	24,678	42.8	51.2	5.1	0.9

NOTE: Percentages may not add to 100 because of rounding.
SOURCE: Bureau of the Census, *Agricultural Economics and Land Ownership Survey* (1988), volume 3, 1987, Census of Agriculture, AG87-R5-2, January 1991.

amount of cash per acre was paid by the farmer to the landowner (table 2–3). Share leases constituted 29.8 percent of the total acreage under lease while other arrangements constituted 4.9 percent. The extent of share leasing differs substantially among regions and states—share leasing is highest in the Midwest, especially in Illinois (61.3 percent).

In general, farmers who lease much of the land they operate can have higher debt-to-asset ratios than owner-operators. Depreciable and current assets make up a higher proportion of the leasing farmer's total assets. Thus, the average debt-to-asset ratio of farm operators, who lease much of the land they operate, tends to exceed that of the aggregate farm sector, as discussed above.

Borrowing

According to the 1988 agricultural finance census survey, an estimated 52.8 percent of the 1.88 million farms in the United States reported the use of debt capital as of December 31, 1988. Thus, nearly half of U.S. farmers made no use of debt at the time of the census survey. Year-end figures could understate the true extent of debt use because some farmers could have repaid short-term operating loans within the year. Among regions, the incidence of indebted farmers was the highest in the Midwest, followed by the West, Northeast, and South. The top three states by proportion of indebted farmers were North Dakota (74.6 percent), South Dakota (70.1 percent), and Iowa (70.0 percent), while the lowest three states were West Virginia (34.9 percent), Tennessee (39.1 percent), and Alabama (39.9 percent). Moreover, the debt was concentrated in larger farming operations. For example, 86.0 percent of the farms with a value of agricultural products sold exceeding $500,000 used debt capital, while fewer than 40 percent of the farms with less than $5,000 of agricultural products sold used debt.

Dispersed Ownership

Agricultural production units in the United States have traditionally been characterized as relatively small operations with a largely noncorporate form of business organization. Ownership, control, management, and risk bearing have generally been concentrated in the hands of individual farmers and their families. These size and concentration characteristics contrast with larger scales of corporate organizations in other business sectors in which ownership, management, and labor are separate functions and in which risk bearing is spread over numerous corporate shareholders.

Within this smaller-scale structure, substantial diversity of farm size occurs. The farm sector is often described as a bimodal distribution of farm sizes: (1) an increasing proportion of large, commercial-scale farms that, while relatively small in number, control most of the sector's wealth, production, and income-generating capacity; (2) many small, part-time and limited-resource farms that rely heavily on nonfarm income; and (3) a decreasing proportion of medium-sized farms. Farms of larger size also use much of the sector's debt capital (table 2–2). (See following discussion about the industrialization of agriculture.)

Riskiness

Agriculture is generally considered a high-risk industry. The traditional business risks come mostly from weather, diseases, and pest infestations affecting production and from unanticipated price variations in resource and commodity markets. The sequential nature of crop and livestock production provides relatively little flexibility in production scheduling. Low price and income elasticities for many commodities subject to weather, international trade conditions, and other uncontrollable events cause wide swings in prices. Moreover, individual farmers have little capacity to control resource or commodity prices. Other uncertainties are associated with personnel performance, tenure security, technological change, and changes in the legal environment of farm businesses. These business risks combine with the financial risks of borrowing and leasing to bring strong challenges in risk management.

Despite the high risks, farmers can employ a broad range of risk management practices (Patrick et al. 1985). Production responses to risk include enterprise diversification, geographic dispersion of farm operations, preventive health and disease practices, and holding of excess machine capacity. Marketing responses to risk include forward contracting, futures and options, inventory management, commodity pools in cooperatives, use of information services, and participation in government price- and income-support programs. Financial responses to risk include holding liquid assets, credit reserves, and insurance and maintaining flexibility in the pace of making new capital investments or replacing depreciable assets. The feasibility of using these methods of risk management varies with the size, type, and other structural characteristics of farm businesses.

Some of these risk management methods are directly influenced by public policies. Commodity programs provide price and income support through target prices and loan rates, and they provide liquidity through nonrecourse loans. Federal crop insurance responds directly to yield risks. Public credit programs administered through the Consolidated Farm Services Agency (formerly called the Farmers Home Administration) of the Department of Agriculture provide credit liquidity as a lender of last resort. Ad hoc disaster programs provide liquidity and other concessions to deal with catastrophic events.

In contrast to the high intrasector risks, a number of studies have shown that rates of return to agricultural assets have low and in some cases negative correlations with rates of return on various types of financial and non-farm assets (Barry 1980; Arthur, Carter, and Abizadeh 1988; Irwin, Forster, and Sherrick 1988; Bjornson 1994; Young and Barry 1987; Christomo and Featherstone 1990). These findings suggest that much of the systematic risk of agricultural investments can be eliminated when they are held in well-diversified portfolios of stocks, bonds, and other real property. Under these conditions, the risk premiums required by investors to hold agricultural investments in well-diversified portfolios would be close to zero. Thus, the high risk of a stand-alone investment in agriculture may be substantially reduced because of the low correlations with a diversified portfolio of assets.

Information Systems

The relatively high risk and smaller-scale structure of the agricultural sector have hampered the development of effective financial reporting systems that convey timely and accurate information about financial performance and creditworthiness to lenders and other interested parties. The result has been a classic case of asymmetric information. This information problem was partially mitigated by the close lender-borrower relationships, the lengthy residency of farmers and their lenders in rural communities, and the related reputation effects. These traditional relationships, however, are changing because of institutional restructuring and consolidation of financial institutions and because of the broadening geographic base of larger farms. In response, financial accounting practices in agriculture are improving significantly as farms become more commercialized, larger, and more reliant on borrowed and leased capital.

Lenders, too, have upgraded the quality of their credit evaluations. Risk-adjusted loan-pricing systems based on more formal credit scoring and risk assessment methods are becoming more widespread (Ellinger, Splett, and Barry 1992). The movements toward greater uniformity in financial reporting by farmers and in credit evaluations are attributable in part to the 1991 recommendations of the Farm Financial Standards Task Force. These developments largely reflect the concerns about loan quality arising from the financial crisis and loan losses of the 1980s, increased competition among lenders, efforts to control lending costs, and improvements in the quality and availability of data from computerized loan information systems.

Most of these advances in information technology involve the larger, commercial-scale farm borrowers. Lack of good information remains the case for smaller, part-time, and limited-resource farms.

Life-Cycle Patterns

The smaller-scale business structure in agricultural production also suggests that the life cycle of the farm operator continues to affect strongly the sector's financial characteristics. As shown in table 2–4, younger farmers initially rely heavily on leasing farmland, then combine ownership and leasing, and finally rely more on their owned land as they approach retirement. Debt-to-asset ratios are highest in the younger age classes and decline steadily as age increases. Farm size and off-farm income also tend to increase with age of operator until ages fifty to sixty and then decline modestly.

Increasing Industrialization

Industrialization of agriculture refers to the increasing consolidation of agricultural production units and to vertical coordination (contracting and integration) among the stages of the food and fiber system. Traditionally, the degree and form of vertical coordination have varied greatly

13

TABLE 2-4
U.S. Farm Financial Data by Age of Operator, 1988

Age of Operator	Number of Farms	% of Farms	% of Debt	Farm Size, Acreage	Acres Rented, %	Acres Owned, %	Off-Farm Income % of farms	Off-Farm Income $ per farm	Operator Interest Paid to Op. Exp.	Average Interest Rates
Under 25 years	26,001	1.4	59.7	340.9	75.2	24.8	64.1	16,862	8.0	9.2
25 to 34 years	219,198	11.7	66.2	363.1	70.4	29.6	78.4	23,463	8.7	8.0
35 to 44 years	346,851	18.5	65.2	418.2	56.6	43.4	80.4	32,490	10.0	8.5
45 to 49 years	207,879	11.1	64.3	447.1	47.7	52.3	78.3	37,525	10.8	8.9
50 to 54 years	213,199	11.3	58.7	502.4	40.0	60.0	75.2	36,785	10.1	9.1
55 to 59 years	233,552	12.4	50.9	492.3	41.4	58.6	74.0	30,074	8.7	9.2
60 to 64 years	222,750	11.9	44.9	517.8	36.0	64.0	71.9	25,481	7.8	9.0
65 to 69 years	178,432	9.5	36.6	473.6	36.0	64.0	72.1	29,586	7.2	8.8
70 years and over	231,705	12.3	27.0	448.0	33.9	66.1	63.2	23,645	5.5	9.2
Average	1,879,567	100.0	52.8	453.4	45.4	54.6	74.4	29,978	9.0	8.8

NOTE: Percentages may not add to 100 because of rounding.
SOURCE: Bureau of the Census, Agricultural Economics and Land Ownership Survey (1988), volume 3, 1987, Census of Agriculture, AG87-R5-2, January 1991.

within agriculture. Nearly all dairy production, seed production, and most commercial fruit and vegetable production for processing in the United States have operated under some form of contract, often involving cooperative organizations. Turkey and egg production rely on both vertical integration and contract production while broilers have been dominated by contract production. In the pork and beef industries, contracting between feeders and processors has grown rapidly in recent years partly because of risk and financing considerations (Rhodes and Grimes 1992). An estimated 15 to 20 percent of fed cattle and hogs is now produced under some form of contract in larger-scale production units. Continued increases are expected.

In the past, government price and income support programs (that is, "contracting" by farmers with the federal government) precluded substantial vertical coordination in grain production although forward contracting of crop sales by farmers was common. Potential reductions in government programs along with processor interest in specific input characteristics will likely lead to more extensive vertical coordination in crop production, especially as input suppliers and processors develop more sophisticated contract alternatives for farmers to consider.

These recent industrialization developments are being fueled by a complex set of consumer, institutional, technological, information, efficiency, financial, and risk-bearing factors (Barkema, Drabenstott, and Welch 1991; Barkema and Cook 1993a, 1993b; Council on Food, Agricultural, and Resource Economics 1994). The effects on the food system's market, financial, and ownership structures are profound. The traditional open market system has increasingly given way to hybrid arrangements with different contractual agreements among suppliers, farmers, processors, distributors, retailers, and consumers and to vertical integration in which two or more stages are controlled by a common entity. These arrangements often extend across international borders.

As a result of these changes, many commercial-scale farmers of tomorrow will be viewed as producers of highly tailored inputs for further processing and packaging rather than as producers of traditional farm products. At the same time, however, some agricultural production will still have a traditional commodity orientation. In these cases, vertical coordination and product differentiation will occur mostly among the processing, distribution, and retail stages. Finally, a mixture of large and small production units will continue, especially when the smaller farms serve specialty markets and have access to off-farm employment. Predicting, understanding, and analyzing the financial effects of these different industrialization models will be challenging.

Summary

The financing characteristics of agriculture suggest a capital-intensive industry in which the dominance of farm real estate has brought liquidity and debt-carrying problems as well as significant reliance on the leasing of farmland by farm operators. Production units are mostly of smaller scale, although the gap is widening between numerous small, part-time, limited-resource farms and the relatively few but much more economically significant commercial-scale operations. Considerable consolidation of production units is occurring, especially in livestock, along with movements toward greater contract production, vertical integration, and financial arrangements with input suppliers and food companies. Reductions in the availability of government contracts through traditional commodity programs will likely bring greater contract opportunities for crop producers with input suppliers and processors. Business and financial risks in agriculture are high, but numerous risk management options are available, especially for larger operations. Clearly, then, questions about the appropriate role of public credit programs in the context of the heterogeneous characteristics of the agricultural sector are important to consider.

3

Farm Debt and Credit Suppliers

As historical data and past analyses show, the use of debt capital in the agricultural sector grew substantially after the World War II era to reach record, yet highly fragile levels in the early 1980s. The annual compound rate of growth for total farm debt increased from 7.1 percent in the 1950s, to 7.9 percent in the 1960s, to 11.7 percent in the 1970s. Between 1975 and 1980, the annual growth rate was 14.4 percent. The breaking of the bubble in the 1980s brought a sharp decline in farm debt that lasted until the 1990s. From a high of $193.8 billion in 1984, total farm debt fell to $137.2 billion in 1989—a drop of 29 percent in five years—before beginning a modest turnaround to reach an estimated $141.4 billion in 1993.

Tables 3–1, 3–2, and 3–3 show the levels and market shares of total farm debt, real estate debt, and non–real estate debt held by the major credit suppliers: the Farm Credit System, commercial banks, life insurance companies, federal lending agencies, individuals, and others. The first four are considered financial institutions because they either specialize in lending or have specialized loan programs for farmers. Individuals and others include agribusinesses and trade firms for non–real estate debt, retiring farmers and other sellers of farm real estate, state credit programs, and other financial institutions with minor involvement in farm lending. These credit suppliers differ substantially in their organizational structures, operating characteristics, sources of funds, and regulatory environment. These differences in turn are reflected in their market shares, competitive positions, and performance in the farm credit market.

Farm Credit System

The Farm Credit System experienced accelerating growth in its farm lending, especially for farm real estate debt, until its share of farm debt peaked at 34 percent in 1982. In the farm real estate market, the FCS share reached nearly 44 percent in 1984. Because of the financial hardships of the 1980s, the FCS share then declined significantly to 1993 levels of 25.2 percent for total farm debt and 32.9 percent for farm real estate debt. The FCS share of non–real estate farm debt was 16.1 percent in 1993, down from a high of nearly 27 percent in 1976.

Commercial Banks

Commercial banks, which were surpassed by the FCS as the dominant farm lender in the 1970s, reestablished their position in the late 1980s. Although commercial banks were also hard hit by the farm financial stresses in the 1980s, the relatively short-term nature and greater diversity of their loan portfolios, along with relatively strong liquidity positions, allowed for a faster recovery than the FCS institutions.

Much recent growth in farm lending by banks is attributed to major expansions in farm real estate lending (or, to be specific, in loans secured by farm real estate). Much of this real estate lending by banks involves adjustable rates in which loans are amortized over fifteen to

19

TABLE 3–1: TOTAL U.S. FARM DEBT, EXCLUDING OPERATOR HOUSEHOLDS, 1976–1993

| | Debt Owed to Reporting Institutions | | | | | | |
	Farm Credit System	Commercial banks	Farmers Home Admin.	Life insurance companies	Total	Individuals and Others[a]	Total Debt
			Millions of Dollars				
1976	29,007	28,077	4,963	6,828	68,874	27,191	96,065
1977	32,992	31,289	6,378	8,150	78,808	32,047	110,855
1978	37,564	34,435	8,833	9,698	90,529	36,871	127,400
1979	45,376	37,125	14,442	11,278	108,222	43,329	151,551
1980	52,974	37,751	17,464	11,998	120,188	46,636	166,824
1981	61,566	38,798	20,802	12,150	133,316	49,065	182,381
1982	64,220	41,890	21,274	11,829	139,214	49,592	188,806
1983	63,710	45,422	21,428	11,668	142,228	48,842	191,070
1984	64,688	47,245	23,262	11,891	147,086	46,701	193,787
1985	56,169	44,470	24,535	11,273	136,447	41,152	177,599
1986	45,909	41,621	24,138	10,377	122,044	34,926	156,970
1987	40,030	41,130	23,553	9,355	114,069	30,342	144,411
1988	37,138	42,706	21,852	9,018	110,714	28,654	139,368
1989	36,218	44,795	18,974	9,045	109,030	28,201	137,231
1990	35,567	47,425	16,950	9,631	109,573	27,794	137,367
1991	35,382	50,169	15,213	9,494	110,259	28,612	138,871
1992	35,616	51,571	13,504	8,718	109,410	29,860	139,270
1993[b]	35,556	53,739	12,211	8,521	110,028	31,327	141,355

Percentage Distribution of Total Debt

Year							Total
1976	30.2	39.2	5.2	7.1	71.7	28.2	100.0
1977	29.8	28.2	5.8	7.4	71.1	28.9	100.0
1978	29.5	27.0	6.9	7.6	71.1	28.9	100.0
1979	29.9	24.5	9.5	7.4	71.4	28.6	100.0
1980	31.8	22.6	10.5	7.2	72.0	28.0	100.0
1981	33.8	21.3	11.4	6.7	73.1	26.9	100.0
1982	34.0	22.2	11.3	6.3	73.7	26.3	100.0
1983	33.3	23.8	11.2	6.1	74.4	25.6	100.0
1984	33.4	24.4	12.0	6.1	75.9	24.1	100.0
1985	31.6	25.0	13.8	6.3	76.8	23.2	100.0
1986	29.2	26.5	15.4	6.6	77.7	22.3	100.0
1987	27.7	28.5	16.3	6.5	79.0	21.0	100.0
1988	26.6	30.6	15.7	6.5	79.5	20.5	100.0
1989	26.4	32.6	13.8	6.6	79.5	20.5	100.0
1990	25.9	34.5	12.3	7.0	79.8	20.2	100.0
1991	25.5	36.1	11.0	6.8	79.7	20.6	100.0
1992	25.6	37.0	9.7	6.3	78.6	21.4	100.0
1993[b]	25.2	38.0	8.6	6.0	77.8	22.2	100.0

NOTE: Percentages may not add to 100 because of rounding.
a. Includes individuals and others (land for contract, merchants and dealers credit, and the like), CCC storage and drying facilities loans, and Farmer Mac loans.
b. Preliminary data.
SOURCE: U.S. Department of Agriculture, *Agricultural Income and Finance, Situation and Outlook Report*, Economic Research Service, AIS-52, February 1994. Washington, D.C.

TABLE 3–2
REAL ESTATE FARM DEBT, EXCLUDING OPERATOR HOUSEHOLDS, 1976–1993

| | Debt Owed to Reporting Institutions | | | | | | CCC | |
	Farm Credit System	Farmers Home Admin.	Life insurance companies	Commercial banks	Total	Individuals and Others[a]	Storage and drying facilities	Total real estate
			Millions of Dollars					
1976	16,881	3,311	6,828	6,075	33,094	17,258	144	50,496
1977	19,640	3,613	8,150	6,994	38,397	19,556	492	58,445
1978	22,686	3,746	9,698	7,717	43,847	21,712	1,148	66,707
1979	27,322	6,254	11,278	7,798	52,653	25,660	1,391	79,704
1980	33,225	7,435	11,998	7,765	60,423	27,813	1,456	89,692
1981	40,298	8,096	12,150	7,584	68,128	29,318	1,342	98,788
1982	43,661	8,298	11,829	7,568	71,357	29,326	1,127	101,810
1983	44,318	8,573	11,668	8,347	72,906	29,388	888	103,182
1984	46,596	9,523	11,891	9,626	77,636	28,438	623	106,697
1985	42,169	9,821	11,273	10,732	73,994	25,775	307	100,076
1986	35,593	9,713	10,377	11,942	67,725	22,660	123	90,408
1987	30,646	9,430	9,355	13,541	62,972	19,380	46	82,398
1988	28,372	8,953	9,018	14,397	60,740	16,873	21	77,634
1989	26,674	8,130	9,045	15,551	59,400	15,939	12	75,351
1990	25,719	7,576	9,631	16,158	59,083	15,047	7	74,137
1991	25,160	7,001	9,494	17,315	58,970	15,623	4	74,597
1992	25,271	6,361	8,718	18,659	59,009	16,628	2	75,639
1993[b]	25,007	5,831	8,521	19,539	58,899	17,116	1	76,016

Percentage Distribution of Debt

Year								
1976	33.4	6.6	13.5	12.0	65.5	34.2	0.3	100.0
1977	33.6	6.2	13.9	12.0	65.7	33.5	0.8	100.0
1978	34.0	5.6	14.5	11.6	65.7	32.5	1.7	100.0
1979	34.3	7.8	14.2	9.8	66.1	32.2	1.7	100.0
1980	37.0	8.3	13.4	8.7	67.4	31.0	1.6	100.0
1981	40.8	8.2	12.3	7.7	69.0	29.7	1.4	100.0
1982	42.9	8.2	11.6	7.4	70.1	28.8	1.1	100.0
1983	43.0	8.3	11.3	8.1	70.7	28.5	0.9	100.0
1984	43.7	8.9	11.1	9.0	72.8	26.7	0.6	100.0
1985	42.1	9.8	11.3	10.7	73.9	25.8	0.3	100.0
1986	39.4	10.7	11.5	13.2	74.8	25.1	0.1	100.0
1987	37.2	11.4	11.4	16.4	76.4	23.5	0.1	100.0
1988	36.5	11.5	11.6	18.5	78.2	21.7	0.0	100.0
1989	35.4	10.8	12.0	20.6	78.8	21.2	0.0	100.0
1990	34.7	10.2	13.0	21.8	79.6	20.3	0.0	100.0
1991	33.7	9.4	12.7	23.2	79.1	20.9	0.0	100.0
1992	33.4	8.4	11.5	24.7	78.7	22.0	0.0	100.0
1993b	32.9	7.7	11.2	25.7	77.5	22.5	0.0	100.0

NOTE: Percentages may not add to 100 because of rounding.
a. Including Farmer Mac loans.
b. Preliminary data.
SOURCE: U.S. Department of Agriculture, *Agricultural Income and Finance, Situation and Outlook Report*, Economic Research Service, 17IS-52, Feb. 1994, Washington, D.C.

TABLE 3–3
NON–REAL ESTATE FARM DEBT, EXCLUDING OPERATOR HOUSEHOLDS, 1976–1993

| | Debt Owed to Reporting Institutions | | | | | | |
	Commercial banks	Farm Credit System	Farmers Home Admin.	Total	Individuals and others	Total Non-Real Estate	CCC Crop Loans
	Millions of Dollars						
1976	22,002	12,127	1,652	35,781	9,789	45,570	936
1977	24,295	13,352	2,764	40,411	11,999	52,410	4,146
1978	26,718	14,878	5,086	46,682	14,011	60,693	4,646
1979	29,327	18,054	8,188	55,569	16,278	71,847	3,714
1980	29,986	19,750	10,029	59,765	17,367	77,132	3,836
1981	31,215	21,268	12,706	65,189	18,404	83,593	6,888
1982	34,322	20,558	12,977	67,857	19,139	86,996	15,204
1983	37,075	19,392	12,855	69,322	18,566	87,888	10,576
1984	37,619	18,092	13,740	69,451	17,640	87,091	8,428
1985	33,738	14,001	14,714	62,453	15,070	77,523	17,598
1986	29,678	10,317	14,425	54,420	12,143	66,563	19,190
1987	27,589	9,384	14,123	51,096	10,916	62,012	15,120
1988	28,309	8,766	12,899	49,974	11,760	61,734	8,902
1989	29,243	9,544	10,843	49,631	12,250	61,881	5,225
1990	31,267	9,848	9,374	50,490	12,740	63,230	4,377
1991	32,854	10,222	8,213	51,289	12,985	64,274	3,579
1992	32,912	10,346	7,143	51,401	13,230	63,631	4,771
1993[a]	34,200	10,549	6,380	51,129	14,210	65,339	4,000

Percentage Distribution of Debt

1976	48.3	26.6	3.6	78.5	21.5	100.0
1977	46.4	25.5	5.3	77.1	22.9	100.0
1978	44.0	24.5	8.4	76.9	23.1	100.0
1979	40.8	25.1	11.4	77.3	22.7	100.0
1980	38.9	25.6	13.0	77.5	22.5	100.0
1981	37.3	25.4	15.2	78.0	22.0	100.0
1982	39.5	23.6	14.9	78.0	22.0	100.0
1983	42.2	22.1	14.6	78.9	21.1	100.0
1984	43.2	20.8	15.8	79.7	20.3	100.0
1985	43.5	18.1	19.0	80.6	19.4	100.0
1986	44.6	15.5	21.7	81.8	18.2	100.0
1987	44.5	15.1	22.8	82.4	17.6	100.0
1988	45.9	14.2	20.9	81.0	19.0	100.0
1989	47.3	15.4	17.5	80.2	19.8	100.0
1990	49.5	15.6	14.8	79.8	20.1	100.0
1991	51.1	15.9	12.8	79.8	20.2	100.0
1992	51.7	16.3	11.2	79.5	20.8	100.0
1993[a]	52.3	16.1	9.8	78.3	21.7	100.0

NOTE: Percentages may not add to 100 because of rounding.
a. Preliminary data.
SOURCE: U.S. Department of Agriculture, *Agricultural Income and Finance, Situation and Outlook Report*, Economic Research Service, AIS-52, February 1994, Washington, D.C.

twenty years with three- to five-year maturities that allow repricing and refinancing at those times. Despite the growth in bank loans secured by farm real estate, the major form of bank lending is still non–real estate debt. Market shares of commercial banks also vary by regions, with lower shares in the Southeast and delta states.

Much of the commercial bank farm debt is provided by smaller agricultural banks, defined as those whose ratio of farm loans to total loans exceeds the national average ratio. As of June 30, 1993, a total of 3,819 agricultural banks (2,722 with less than $50 million in total assets) held $31.6 billion of agricultural loans, or 56.3 percent of total bank loans to agriculture. Nonagricultural banks held 43.7 percent of bank loans to agriculture, of which a rapidly growing 24.1 percent was held by banks with total assets greater than $500 million (table 3–4).

Insurance Companies

Following World War II, life insurance companies were the largest supplier of farm real estate debt. In 1945, they held about one-quarter of all farm real estate loans. Their relative position declined, however, until leveling out in the 11 to 13 percent range after 1980. Currently, only six life insurance companies are significantly involved in farm real estate lending, mostly with larger loans and with regional shifts away from the Cornbelt and toward the Southeast and Pacific Coast regions (Koenig and Stam 1992).

Agribusiness and Individuals

Most of the non–real estate farm debt provided by individuals and others comes from agribusinesses and trade firms that provide credit services along with merchandising activities. Trade credit, accounting for 20

TABLE 3–4
AGRICULTURAL LENDING OF AGRICULTURAL AND NONAGRICULTURAL BANKS BY BANK SIZE, 1993
(millions of dollars)

	Agricultural Banks					Nonagricultural Banks				
	Number of banks	Total ag. loans ($)	Avg. ag. loans ($)	Ag. lending share[a] (%)	Ag. loans/ total loans (%)	Number of banks	Total ag. loans ($)	Avg. ag. loans ($)	Ag. lending share[a] (%)	Ag. loan/ total loans (%)
Under 25	1,478	5,510	3.7	9.8	47.9	907	402	0.4	0.7	5.0
25–50	1,244	9,272	7.5	16.5	41.7	1,643	1,244	0.8	2.2	3.8
50–100	829	10,037	12.1	17.9	35.8	1,943	2,729	1.4	4.9	3.5
100–300	256	5,940	23.2	10.6	30.7	1,846	4,884	2.6	8.7	2.8
300–500	9	513	57.0	0.9	26.4	371	1,745	4.7	3.1	2.1
Over 500	3	343	114.2	0.6	19.5	611	13,544	22.2	24.1	0.8
Total	3,819	31,615	8.3	56.3	37.3	7,321	24,548	3.4	43.7	1.2

NOTE: Figures are weighted within size class.

a. This figure represents the percentage of total commercial bank agricultural loans held by this size group of banks.

SOURCE: U.S. Department of Agriculture. Agricultural Income and Finance, Economic Research Service, AIS-52, Feb. 1994, Washington, D.C.

27

to 25 percent of non–real estate farm debt since 1975, is especially strong for farm machinery and equipment. Some larger input suppliers have also developed operating credit programs for their farm customers, although total loan volume is relatively small.

Farm real estate debt from individuals mostly involves contracts and mortgages held by sellers of farmland. Seller financing remains substantial ($17.1 billion and 22.5 percent of total farm real estate debt in 1993), although declining significantly from pre-1980 levels when capital gains tax advantages, level of tax rates, and number of tax brackets were substantially reduced.

The Federal Government

Farm lending by the federal government takes several forms. One form is the direct and guaranteed lending by the Farmers Home Administration. FmHA lending has fluctuated inversely with the financial performance of the agricultural sector. It increased sharply during the 1980s, reflecting various types of emergency loan programs and the serious economic problems of agriculture. In fact, FmHA held nearly 23 percent of outstanding non–real estate farm debt in 1987. This share had declined to less than 10 percent by 1993, in part because of a major shift by FmHA from direct loans to guaranteed loans made by commercial lenders (see chapter 6 on FmHA loan programs).

FmHA's involvement in real estate debt is more consistent than in non–real estate debt, and has been in the 6 to 12 percent range since the mid-1970s. The comparatively minor amounts of agricultural lending by state credit programs and the Small Business Administration (in the late 1970s and early 1980s) is included in the categories of individuals and other.

Another form of farm lending by the federal government is nonrecourse price support loans made by the

Commodity Credit Corporation as part of the government's price and income support policies for farmers. In general, CCC price support loans are low when market prices of farm commodities are well above their support levels; when market prices decline to approach the support level, CCC loan volume may increase substantially. CCC loan volume declined to about $4 billion in 1993, or 5.8 percent of total non–real estate debt, from a high of 22.4 percent in 1987. The share of U.S. government non–real estate debt relative to total non–real estate farm debt reached a high of 39.2 percent in 1987, when FmHA and CCC loans are combined.

4

Farm Credit Programs and Financial Policies

Evaluating the relationships among credit markets, the agricultural sector, and public programs, policies, and subsidies is complicated by the multiple perspectives involved. These perspectives include (1) credit as a key source of financial capital for agricultural production; (2) credit and the changing structure of the agricultural sector; (3) credit markets and the concept of a "level playing field" for the major providers of agricultural credit; and (4) credit as a political tool.

Credit as Financial Capital

Many farmers have relied heavily on credit to finance their capital base, to mechanize and modernize their farming operations, to conduct marketing and production plans, and to serve as a valuable source of liquidity in responding to risks. Many significant, long-term changes in the farm sector—larger farm size, fewer farm numbers, greater specialization, greater capital intensity, adoption of new technology, stronger market coordination—have been facilitated by readily available credit. Thus, equitable access to credit is important to the economic contributions and financial performance of the agricultural sector.

Structure of the Agricultural Sector

Despite these structural changes and the related roles of credit, the historical developments in public credit for agriculture have been important in maintaining a pluralistic, smaller-scale, largely (but by no means completely) noncorporate organization of agricultural production units, seemingly consistent with the public interest. Credit has been viewed as a facilitating tool for aggregate structural change but not as a driving force (Gustafson and Barry 1993). The increasing industrialization of agriculture, however, especially in livestock and poultry production, is challenging this historical perspective.

A Level Playing Field

From a regulatory point of view, the intangible nature of financial assets and services, along with the need for confidence, trust, and stability among financial market participants, brings considerable government regulation of financial institutions. The extent of regulation varies substantially among credit sources, ranging from comprehensive oversight of depository institutions (including agricultural banks) to specialized government-sponsored enterprises to the largely unregulated lending by agribusinesses and individuals. This regulatory mosaic can create periodic imbalances in competition in credit markets that raise concerns by the participants about leveling the regulatory playing field. The recent deregulation of depository institutions, including the current focus on interstate banking, and the restructuring of the Farm Credit System have clouded the field and heightened interest in the regulatory balance issue.

31

Credit as a Political Tool

Finally, from a policy maker's perspective, credit programs are a popular, politically expedient policy instrument (Barry 1985; Barry and Boehlje 1986). They are relatively easy and cost effective to administer, as long as program demands are not growing too fast. While the administrative and risk-bearing costs are difficult to measure and are usually hidden from taxpayers, the programs are highly visible to constituents. They can be targeted to selected groups, they can be quickly developed for responding to ad hoc crises, and they do not directly influence commodity and input markets, although the secondary effects on asset values, income, and risk can be significant. Moreover, credit programs give the impression of financial soundness because loan repayment with interest is intended.

These differing perspectives occur concurrently and can create trade-offs in goals and purposes of public credit programs and financial policies. Credit policies intended to maintain the pluralistic structure of the agricultural sector can slow resource adjustment, build excess production capacity, and counter the effects of new technologies and market forces in the farm sector (Lee and Gabriel 1980). Emergency or disaster-related public credit can in effect substitute credit for income, thus perpetuating perverse incentives. Weak monitoring and enforcement problems in public credit can create moral hazards by borrowers who prefer to maintain their eligibility for the program (LaDue 1990). Private lenders may even take excessive lending risks if they can count on public credit as a reliable safety net. These actions and effects may in turn undermine the integrity and soundness of the credit markets. They complicate the effective design and performance evaluations of public credit programs.

5
Public Credit

The rapid growth in public credit programs and their prominent position in the financial markets have brought increasing attention to the intended goals, mechanisms, subsidies, payoffs, accounting procedures, and general effectiveness of these programs (Bosworth, Carron, and Rhyne 1987; Budget of the United States 1994). The attention is broadly focused on all types of public credit—for housing, small businesses, education, and agriculture—and on each of the major forms of public credit: (1) direct loans; (2) guaranteed loans; (3) government-sponsored enterprises; and (4) tax-exempt state and local bonds (see Bosworth, Carron, and Rhyne 1987; Gale 1991; and Office of Management and Budget "Special Analysis F" for comprehensive discussions of these programs). Especially important are understanding and evaluating the basic missions of these programs.

Missions and Types of Programs

Public credit programs are intended either to correct an imperfection, to fill a gap in the workings of credit markets, or to achieve a public purpose through the reallocation of resources or redistribution of income in the economy. Programs that aim to correct market imperfections need not require subsidization; they are considered the more successful government programs in credit markets (Bosworth, Carron, and Rhyne 1987). In contrast, efforts to achieve public purposes generally do involve subsidization, with significant questions raised about the form, magnitude, length, measurability, and recipients of the subsidies.

Major questions also concern the appropriateness of credit programs relative to other mechanisms for providing the subsidy. Credit programs have weaknesses in transmitting subsidies because the loan funds may be used for unintended purposes, the borrowers may have had access to credit from other sources, the subsidy benefits may accrue to private lenders rather than to borrowers, or favorable terms of credit may be capitalized into the values of the assets being financed. Moreover, using credit markets to transmit subsidies undermines the integrity of inherently fragile financial markets. A financial market's primary function is to facilitate financial intermediation by adjusting the liquidity and risk positions of savers and investors. Because credit transactions involve intangible financial assets and promises to repay, high levels of confidence, trust, discipline, and stability are needed for these markets to function effectively. Extensive government regulation contributes to market effectiveness. Adding a subsidy role, however, is counterproductive. Thus, the larger the subsidy needed to achieve the public purpose, the less the assistance should be channeled through public credit programs.

Among the forms of credit programs, the emphasis has clearly shifted toward guaranteed loans and away from direct loans. Loan guarantees have been shown to provide lower subsidies than direct loans, especially since direct loans are seldom priced to cover the government's full cost of funding, administering, and risk bearing. Pricing for risk through fees or premiums is more explicit with a loan guarantee. Loan guarantees also displace fewer financial market resources, offer greater liquidity for loan sales, and provide greater use of private lenders' knowledge and experience for loan origination, servicing, and management. Disadvantages of loan guarantees are loss of direct loan control and less capacity to respond for specific borrower groups and events. Among the federal credit programs, direct lending has been most prominent in farm lending by the Farmers Home Administration, although FmHA programs have recently swung substantially toward guarantees.

Government-sponsored enterprises are privately owned financial institutions chartered and established by the federal government to serve the credit needs for agriculture, housing, and college students. These borrowing groups were considered unserved or underserved by existing credit markets when the institutions were created. Currently included among those enterprises are the Farm Credit System, the Federal Agricultural Mortgage Corporation (Farmer Mac), Student Loan Marketing Association (Sallie Mae), the Federal Home Loan Bank System, the Federal National Mortgage Association (Fannie Mae), the Federal Home Loan Mortgage Corporation (Freddie Mac), and the College Construction Loan Insurance Association.

To aid with their targeted lending programs and related concentrations of lending risks, the government-sponsored enterprises were given various regulatory preferences and exemptions in their access to the financial markets (see chapter 7 in this volume). Moreover, investors in the debt securities of those enterprises have had the widespread perception that the federal government will back these securities in stressful times, even though the government has no legal obligation to do so. Thus, funding costs for such enterprises are generally less than those not backed by the government. In addition, the agricultural enterprises sponsored by the government and the Federal Home Loan Banks each have specific government regulators to monitor risk taking, enforce capital requirements, and oversee the mandated public purposes. The other government-sponsored enterprises are supervised by broader-based government agencies and departments.

The Farm Credit System is unique among the government-sponsored enterprises in at least two other respects. First, the FCS concentrates on making direct loans through a network of banks and lending associations to eligible agricultural borrowers. Other such enterprises focus primarily on maintaining secondary markets for reselling mortgages and student loans, offering secondary loan guarantees, and providing credit to private lending institutions (Bosworth, Carron, and Rhyne 1987). Second, the institutions of the FCS are organized as cooperatives in which the agricultural borrowers become the owners of the system and are represented by elected boards of directors. Thus, the equity capital, ownership direction, and risk bearing of the FCS are concentrated in the hands of its borrowers. Other government-sponsored enterprises with broader-based ownership experience greater diversity in financial oversight by owners and in the aggregate capacity of ownership to bear risks.

Management, Control, and Subsidy Measurement

Since the early 1980s, the federal government has substantially clarified its goals and upgraded its procedures for managing, controlling, and accounting for federal credit programs. In 1984, the Office of Management and Budget issued guidelines for a more systematic budget and accounting process that clearly shows the magnitude, composition, and form of federal credit and estimates the subsidies conveyed to borrowers. The OMB also produced new guidelines for agencies to use in program administration. These guidelines stipulate that existing and new credit programs will experience rigorous annual reviews of objectives, stronger justification, explicit statements of the borrower's subsidies, pricing policies related to market interest rates and insurance premiums, a preference for loan guarantees, and private sector sharing of risk-bearing costs to the extent possible. The individual agencies are left to carry out these guidelines.

The Federal Credit Reform Act of 1990 fundamentally changed the budgetary treatment of direct loans and loan guarantees (Budget of the United States 1994, 144). The act, which became effective in 1992, requires budgeting for the "costs" of federal credit programs. These costs are defined as "the estimated long-term cost to the government of a direct loan or a loan guarantee, calculated on a net present value basis, excluding administrative costs." Before the passage of the act, direct loans and loan guarantees were recorded on a cash basis. Direct loans were shown as a cash outlay in the year of disbursement, and repayments were recorded as they occurred over a period of years. Loan guarantees were recorded in the federal budget only when fees were received by the government or when the government made payments for default claims. This past system tended to overstate the cost of direct loans when they were made and understated the cost of loan guarantees. Several effects occurred (Budget of the United States 1994, 144):

- The federal government was encouraged to favor loan guarantees over direct loans regardless of their true costs.
- The true costs of different credit transactions could not be compared with one another or with other budgetary programs such as grants.
- Direct loans and loan guarantees affected budget outlays in later years, after their costs were sunk and largely uncontrollable.
- Payments for guarantee claims did not require the discipline of normal appropriations.
- Reserves were generally not set aside for loan guarantees to pay for probable defaults.

The 1990 act responded to these shortcomings by recording the full costs (subsidy) as an obligation when the government enters into a loan obligation or guarantee commitment. The full cost of the loan or guarantee is recorded as a budget outlay when the direct loan or guaranteed loan is disbursed to the public. The subsidy element of a credit program is calculated as the difference between the present value of the expected cash outflows from the government and the present value of the expected cash inflows, each discounted by the interest rate on marketable Treasury securities of like maturity at the time of loan disbursement. Reestimates of the subsidy cost of a government program then occur periodically throughout the lifetime of a loan or guarantee.

This new cost-based focus of accounting for federal credit programs provides more accurate information for decision making and for making interprogram comparisons of lending costs. It also allows comparisons among the costs of government assistance to different sectors of the economy. Cost effects of modifications of credit programs can also be evaluated. The cost-based approach to credit accounting, however, is not the same as measuring the value of the subsidy benefit to borrowers. Borrower subsidies are based on comparisons of costs of federal credit programs with those of other sources of financing. A current perspective on the magnitude, costs, and subsidy rates of federal credit programs in fiscal year 1995 is provided in tables 5–1, 5–2, and 5–3. Table 5–1 indicates the face values and cost estimates of federal direct loans, guaranteed loans, federal insurance, and government-sponsored enterprises for 1992 and 1993. Tables 5–2 and 5–3 indicate the estimated 1995 subsidy rates and new lending authorities for various categories of direct and guaranteed loans. The entries for the Agricultural Credit Insurance Fund represent the farm lending programs of FmHA. The estimated subsidy rates for these programs are 13.03 percent for direct loans and 2.49 percent for guaranteed loans. In general, the subsidies on direct loans exceed those on guaranteed loans by relatively large margins.

New Theories of Public Credit

As indicated in the preceding section, public credit programs are generally rationalized in terms of market imperfections or achievement of public purposes through the reallocation of resources or redistribution of income. Recent theoretical work (Mankiw 1986; Gale 1987, 1990a, 1990b, 1991; Smith and Stutzer 1989) has focused on informational imperfections in credit markets as a motivation for federal credit programs to enhance market performance. These developments have substantially enhanced the theoretical foundations of public credit programs as responses to informational imperfections.

This new work builds on earlier efforts by Jaffee and Russell (1976), Stiglitz and Weiss (1981), and others to explain credit rationing as a rational, equilibrium-generating response to adverse-selection problems attributable to asymmetric information between lenders and borrowers. That is, when lenders are unable to distinguish between high- and low-risk borrowers on loan projects or when borrowers within high- and low-risk classes cannot be distinguished, the practice of charging an average interest rate leads low-risk borrowers to self-finance or seek another lender and leaves the original lender with only high-risk borrowers. When such adverse selection is left to run its course, the ultimate risk is market failure.

High-risk borrowers present special problems because they are believed to select the high-rate loans and to undertake riskier actions as interest rates increase. Thus, higher interest rates do not ensure higher profits to lenders because of increasing risk-bearing costs as the loan rates increase. In general, then, changes in interest rates alter the riskiness of the pool of borrowers.

To avoid this adverse-selection problem, the rational lender can ration credit through rejection of seemingly

TABLE 5-1
FACE VALUE AND ESTIMATED COST OF FEDERAL CREDIT AND INSURANCE PROGRAMS, 1992–1999
(billions of dollars)

Program	Face Value 1992[a]	1994 Estimates of Future Costs[a,b]	Face Value 1993	Current Estimates of Future Costs[b]	Subsidy Outlays 1994–99
Direct loans[c]					
Farm Service Agency Rural Development Administration	50	16–22	49	18–24	4–6
Rural Electrification Administration and Rural Telephone Bank	38	2–4	36	3–5	1–2
Federal Direct Student Loan Program	—	—	—	7–10	3–5
Export-Import Bank	9	3–5	9	3–5	0–1
Agency for International Development	16	5–7	14	5–7	0–1
Public Law 480	12	7–9	12	7–9	2–3
Foreign military financing	9	2–3	9	0–2	0–1
Small business	6	2–3	6	2–3	0–2
Other direct	16	2–4	16	2–4	0–1
Total direct loans	156	39–57	151	47–69	10–22
Guaranteed loans[c]					
FHA single-family	308	(14)–0	292	(18)–0	(10)–0
VA mortgage	176	3–6	161	3–6	1–3

FHA multifamily	79	3–6	81	4–6	0–1
Federal Family Education Loan Program	79	20–30	85	8–11	7–9
Small business	17	1–3	20	2–4	1–3
Farm Service Agency	6	1–3	7	1–4	0–1
Export-Import Bank	8	4–7	12	4–5	1–3
CCC export credits	9	4–5	9	4–5	2–3
Other guaranteed	22	0–1	26	1–3	0–1
Total guaranteed loans	704	22–61	693	9–44	2–24
Federal insurance					
Banks[d]	1,943	1–12	1,889	30–45	15–30
Thrifts[d]	761	25–37	707	15–25	5–15
Credit unions	218	—	237	—	—
Total deposit insurance	2,922	26–49	2,833	45–70	20–45
PBGC[d]	950	25–40	950	60–90	18–20
Disaster insurance	721	7–9	722	10–16	9–13
Other insurance	358	4–6	511	9–10	8–9
Total federal insurance	4,951	62–104	5,016	124–186	55–87
GSEs[e]					
Freddie Mac	427	—	474	—	—
Fannie Mae	543	—	622	—	—
Federal Home Loan Banks	85	—	107	—	—

(Table continues)

TABLE 5-1 (continued)

Program	Face Value 1992[a]	1994 Estimates of Future Costs[a,b]	Face Value 1993	Current Estimates of Future Costs[b]	Subsidy Outlays 1994-99
Sallie Mae[f]	—	—	—	—	—
Farm Credit System	50	0-1	52	0-1	0-1
Total GSEs	1,105	0-1	1,255	0-1	0-1
Total	6,916	123-223	7,115	180-300	67-134

a. Costs are as they were displayed in the 1994 budget, uncorrected for errors; face values for 1992 have been updated.
b. Direct loan future costs are program account outlays projected into the future plus the embedded loss from outstanding loans. Loan guarantee costs are program account outlays plus liquidating account outlays (and outlays from defaulted guarantees that result in loans receivable) projected into the future. Future insurance costs are the equivalent of program plus liquidating costs through 1996, plus the accrued liability remaining at the end of 1996.
c. Excludes loans and guarantees by deposit insurance agencies and programs not included under credit reform, such as CCC farm supports. Defaulted guarantees that become loans receivable are accounted for in guaranteed loans.
d. Current estimates of deposit insurance and pension insurance costs reflect improvements in estimation methods. The corrected estimated costs for deposit insurance in 1994 were $80-90 billion (banks: $35-53 billion; thrifts: $25-37 billion), while pension insurance costs were $45-75 billion.
e. Net of borrowing from federal sources, other GSEs, and federally guaranteed loans.
f. The face value and federal costs of guaranteed student loans in Sallie Mae's portfolio are included in the education account above.

Source: Budget of the United States Government, *Analytical Perspectives FY 1995*, U.S. Government Printing Office, Washington, D.C., 1994.

TABLE 5-2
ESTIMATED 1995 SUBSIDY RATES, BUDGET AUTHORITY, AND LOAN LEVELS FOR DIRECT LOANS
(millions of dollars)

Agency and Program	1995 Weighted Average Subsidy as a Percentage of Disbursements	1995 Subsidy Budget Authority	1995 Estimated Loan Levels
Funds appropriated to the president			
Micro and small enterprise development	8.10	[a]	1
Regional peace and security (formerly FMF)	7.74	60	770
Overseas Private Investment Corporation	14.22	3	20
Agriculture			
Agricultural credit insurance fund	13.30	125	937
Rural telecommunication partnership loans	4.24	1	15
Rural housing and community development service	9.50	28	300
Self-help housing	2.99	[a]	[a]
Rural housing insurance fund	14.03	315	2,248
Rural development loans	52.25	65	125
Rural economic development loans	23.92	3	13
Rural electric and telephone	1.40	19	1,354

(Table continues)

TABLE 5–2 (continued)

Agency and Program	1995 Weighted Average Subsidy as a Percentage of Disbursements	1995 Subsidy Budget Authority	1995 Estimated Loan Levels
Rural electric and telephone refinancing (mandatory)	16.85	13	1,500
Rural telephone bank	0.02	[a]	175
Rural utilities service	13.97	136	977
Public Law 480 direct loans	81.06	278	340
Education			
Federal direct student loan program	7.32	349	4,765
Housing and Urban Development			
FHA-mutual mortgage insurance direct loans	—	—	180
FHA-general and special risk direct loans	—	—	220
Interior			
Bureau of Reclamation loans	27.49	3	11
State Department of Repatriation loans	80.00	1	1
Transportation			
Minority business resource center program	10.00	2	15

Veterans Affairs			
Transitional housing loans	10.00		a
Direct loan	11.76		a
Loan guarantee fund	2.34	18	783
Guaranty and indemnity fund	1.06	6	554
Education loan fund	26.47	a	a
Vocational rehabilitation	2.80	a	2
Other Independent Agencies			
Community development financial institutions fund	13.19	20	152
Export-Import Bank[b]	17.91	371	2,070
Federal Emergency Management Agency			
Disaster Assistance	9.67	2	25
Small Business Administration			
Business loans	2.99	—	—
Disaster loans	12.67	52	412
Total	—	1,869	16,464

NOTE: Additional information on credit reform subsidy rates is contained in the Federal Credit and Insurance Supplement to the budget for 1995.

a. $500,000 or less.

b. Includes FY 1993 and 1994 carryover budget authority.

SOURCE: Budget of the United States Government, *Analytical Perspectives FY 1995*, U.S. Government Printing Office, Washington, D.C., 1994.

TABLE 5-3
ESTIMATED 1995 SUBSIDY RATES, BUDGET AUTHORITY, AND LOAN LEVELS FOR LOAN GUARANTEES
(millions of dollars)

Agency and Program	1995 Weighted Average Subsidy as a Percentage of Disbursements	1995 Subsidy Budget Authority	1995 Estimated Loan Levels
Funds appropriated to the president			
Micro and small enterprise development	5.46	1	26
AID housing and other credit guarantees	14.55	12	82
Overseas Private Investment Corporation	1.83	9	482
Agriculture			
Agricultural credit insurance fund	2.49	72	2,879
Agricultural resource conservation demonstration	56.10	3	6
CCC, export credits	6.92	394	5,700
Rural housing and community development service	4.97	4	75
Rural housing insurance fund	1.72	22	1,300
Rural business and cooperative development service	0.95	11	1,116
Commerce			
Economic development guarantees	18.56	50	269

Education			
Federal family education loan program	11.26	1,844	16,382
Health and Human Services			
Health professions graduate student loan program	6.74	25	375
Housing and Urban Development			
Community development (Sec. 108)	—	—	2,054
Federal Housing Administration general and special risk[a]	3.73	152	19,685
Federal Housing Administration mutual mortgage	-2.78	—	84,962
GNMA secondary mortgage guarantees	—	—	130,000
Commodity development loan guarantees	—	—	2,054
Interior			
Indian loan guaranty and insurance fund	18.09	10	47
Transportation			
Title XI maritime guaranteed loans	9.88	50	500
Veterans Affairs			
Guaranty and indemnity fund	1.18	357	30,256
Loan guaranty fund	13.34	[b]	2
Other independent agencies			
Export-Import Bank[c]	4.34	675	15,565

(Table continues)

TABLE 5–3 (continued)

Agency and Program	1995 Weighted Average Subsidy as a Percentage of Disbursements	1995 Subsidy Budget Authority	1995 Estimated Loan Levels
Small Business Administration business loans	2.99	318	11,419
Total	—	4,009	325,255

NOTE: Additional information on credit reform subsidy rates is contained in the Federal Credit and Insurance Supplement to the budget Year 1995.
a. Subsidy rate shown is for positive subsidy risk categories only.
b. $500,000 or less.
c. Includes FY 1993 and 1994 carryover budget authority.

eligible borrowers (Stiglitz and Weiss 1981) or limit the availability of credit on seemingly eligible loan projects (Gale and Hellwig 1985). These rationed borrowers or projects encounter this misfortune because of the lender's inability to distinguish creditworthiness or the borrower's inability to signal its creditworthiness.

Mankiw (1986, 469) shows how government credit programs that allocate credit among competing uses can, under plausible conditions, improve on the "unfettered" market equilibrium. He observes: "A necessary condition for efficient government intervention is unobservable heterogeneity among would-be borrowers regarding the probability of default. The greater is such heterogeneity, the greater is the potential for efficient intervention," even if the government has no informational advantage over private lenders.

Mankiw's analysis shows that, under asymmetric information about the probability of loan repayment, low-probability investments are overly encouraged and high-probability investments are overly discouraged. A gain in efficiency may then occur if intervention by direct or guaranteed loans reduces the loan rate toward a marketwide risk-free rate so that a reduction occurs in rationing of credit to borrowers with potentially profitable investments. The rate reduction allows some of those borrowers with high returns and high repayment probabilities who were not previously investing to invest now, but it also induces some borrowers with low returns and low repayment probabilities to invest as well. Ultimately, a successful credit program must have marginal benefits exceeding the marginal dead-weight loss of funding the program through distortionary taxation.

Mankiw also shows how small changes in the risk-free, required rate of return by lenders can cause large and inefficient changes in credit allocation and even market collapse in some cases. The high social cost of a market collapse triggered by informational problems and rising loan rates could "reasonably motivate the govern-

FIGURE 5–1
LENDER'S EXPECTED RETURNS ON GENERAL LOANS (ρ_G) AND TARGET GROUP LOANS (ρ_j)

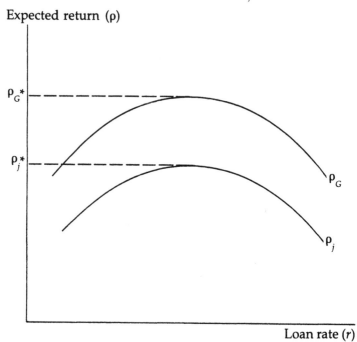

Expected return (ρ)

ρ_G^*

ρ_j^*

ρ_G

ρ_j

Loan rate (r)

ment to act as a lender of last resort" (p. 468).

Gale's theoretical analysis (1990a, 1990b, 1991) poses a competitive market case of asymmetric information in which risk-neutral lenders can identify groups of risk-neutral borrowers, some of which may be candidates for government credit policies, but the lenders cannot identify differences in risk within groups. He establishes the assumptions that, first, loan demand is a decreasing function of the effective interest rate (net of credit policy effects) and, second, repayment risk increases as the effective rate increases, r_i^* (resulting in adverse selection). Given these assumptions, an increase in gross lending rates, r_i, eventually causes a decline in the lender's expected return (ρ) from lending, and the specific response

FIGURE 5–2

SUPPLY AND DEMAND FOR LOANS, ALTERNATIVE BORROWER GROUPS

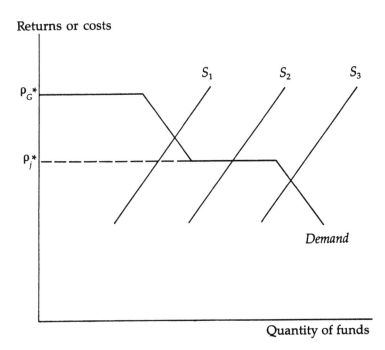

patterns may differ among the borrower groups (see return patterns of the target group and the general group in figure 5–1) (Gale 1991, 137). Then, the addition of the plausible assumption that the lender's maximum expected return on general loans is at least as great as the maximum expected returns for groups targeted for credit policies (that is, $\rho_G^* \, \rho_j^*$), establishes the conditions for situations in which the general market clears while the target groups are rationed (partially financed) or red-lined (excluded from financing).

These conditions occur as lenders order the borrower groups according to their maximum expected returns, relative to equilibrium funding costs, and finance the groups sequentially. This situation is described by the step function for effective loan demand in figure 5–2 (Gale 1991,

138). The market will clear for all groups in which maximum expected returns (ρ_G^*) from general lending exceed the equilibrium cost of funds ($\hat{\rho}$)—supply function S_1 in figure 5–2. Rationing will occur for groups with maximum expected returns equal to the equilibrium rate ($\rho_j^* = \hat{\rho}$)—S_2 in figure 5–2. And red-lining will occur for groups with maximum expected rate of return less than the equilibrium cost of funds—($\rho_j^* < \hat{\rho}$)—S_3. Consequently, the marginal effects of changes in supply and demand for funds fall completely on the rationed or red-lined groups that are targeted for credit policies.

Given these conditions, credit policies (subsidized lending, loan guarantees, tax exempt lending) then may act directly on the equilibrium positions by creating a wedge between loan rates and effective rate levels. The greater the wedge, the lower the effective loan rate is. In turn, lower effective loan rates will reduce repayment risks and increase expected returns from lending. These increases in expected returns to lenders for the respective target groups will help resolve the rationing and red-lining problems. Gale elaborates as follows:

> In the presence of rationing, pure interest subsidies do little to assist targeted borrowers. The interest subsidy operates primarily by reducing borrower payments; however, rationing exists because of insufficient credit worthiness of borrowers, not their unwillingness to pay. In contrast, loan guarantees operate equally effectively in either clearing or rationed regimes, because they operate primarily by raising the lender's return. These results indicate that guarantees are relatively more effective than subsidies in reallocating credit if the target group is rationed or redlined than when the target market clears (1990a, 190).

Gale uses this theoretical framework to evaluate the

allocational effects of federal credit programs existing in the 1980s and their efficiency costs caused by the government's own financing needs. He uses a variety of numerical estimates and assumptions about the form and extent of subsidies, investment demand elasticities, repayment rates, supply elasticities, and other model parameters to investigate these issues.

According to Gale's results, credit policies have important allocational effects, due to the large subsidies, and substantial efficiency costs, attributed to high government financing costs. Existing credit subsidies were estimated to raise aggregate private investment by between 0 and 4 percent depending on the supply elasticity for funds. The allocational effects depend on the size of the effective subsidy rather than on credit volume. Thus, high-subsidy, low-volume programs have greater allocational effects than their counterparts.

The estimated efficiency cost of credit policy is high, ranging from $10 billion to $15 billion, or about one-third of 1 percent of GNP in 1987—or in excess of fifty cents per dollar of incremental target group investment. The latter cost reflects the tendency for most direct welfare gains to go to borrowers who would have received credit without government subsidies. For most of Gale's simulations, the gains and losses to the various borrower groups and suppliers of funds roughly offset each other. Given these offsets, and the high financing costs, the credit programs of the 1980s appear to require large external benefits to be welfare improving. Finally, interactions among credit programs can indirectly eliminate much or all of the original program's direct gain.

Smith and Stutzer (1989) also use the asymmetric information condition to develop a theoretical model that suggests the rationing of low-risk borrowers so that high-risk borrowers will not misrepresent themselves to obtain the preferred terms of the low-risk loan contract. In this case, rationing arises because of the lender's inability to partition borrowers into risk classes ex ante accurately and cost effectively. A loan-guarantee program, available for all types of loans, is shown to reduce interest rates for both high- and low-risk borrowers and to increase the low-risk borrower's probability of receiving a loan. These effects are claimed to increase social efficiency, although the government's cost of funding the program is not considered. In contrast, a direct loan program increases the adverse-selection problem by making it more desirable for high-risk borrowers to misrepresent their type. The result then is a trade-off of either higher social efficiency or decreased welfare of the rationed borrowers.

In commenting on Smith and Stutzer's paper, Penner (1989) points out a number of anomalies between the theoretical analysis and the provisions of actual credit programs. Especially important are the practice of limiting guarantee programs to targeted groups of borrowers, the percentages of losses typically guaranteed, moral-hazard actions by governments themselves to prevent guaranteed loans from going bad, the roles of government-sponsored enterprises, and the tendency for rationing to encourage higher-risk rather than lower-risk borrowers.

6
The Farmers Home Administration

The period from the early 1900s through the 1930s witnessed the basic shaping of today's agricultural credit markets.[1]

Foundations and Evolution of Agricultural Credit

Some developments included institutional reforms and innovations in private credit markets, especially in commercial banking and the Farm Credit System. Developments affecting commercial banking were the creation of the Federal Reserve System in 1913 and a series of legislative enactments that clearly delineated the uniqueness of banking (the Glass-Steagall Act of 1933), allowed the diverse geographic structure of banking (the McFadden Act of 1927), and established control over the pricing and risk environment of banking in the 1930s through ceilings on interest rates (removed during the 1980s), portfolio requirements, legal lending limits, and federal deposit insurance. These developments stabilized the banking system and helped sustain the significant involvement in many states of small rural banks in financing agriculture.

1. This brief review of the historical foundations and evolution of the agricultural credit markets is drawn from Barry and Boehlje (1986).

For the FCS, the period up to 1935 saw the development of federal land banks starting in 1916, followed by the federal intermediate credit banks in 1923, and production credit associations and banks for cooperatives in 1933. Thus, the basic framework of the FCS was completed during the 1930s. The long-term goal was to move the system toward a private status in which ownership and control would rest with the financial institutions and ultimate agricultural borrowers, funding would occur from the private financial markets (with agency status of the system's debt securities), and government control would be limited to the regulatory and supervisory functions experienced by other types of federally chartered financial institutions. These goals were largely accomplished, although significant consolidations and restructurings of FCS institutions occurred as a result of financial stresses and legislative enactments of the 1980s.

Since the 1930s, the FCS mission, as stated in the Farm Credit Act of 1971 (as amended) has been to "accomplish the objective of improving the income and well-being of American farmers and ranchers by furnishing sound, adequate, and constructive credit and closely related services. . .necessary for efficient farm operations" and "a permanent system of credit for agriculture which will be responsive to the credit needs of all types of agricultural producers *having a basis for credit* [emphasis added]".

These early developments in the twentieth century set the broad outlines for providing loan funds from *commercial sources* to *creditworthy* agricultural borrowers throughout the United States. Providing financing to agricultural borrowers who could *not* obtain credit from commercial sources, however, was another policy issue. The characteristics of those borrowers have changed over time but have generally included impoverished, destitute farm families, young farmers entering the sector, small yet potentially viable farms, limited-resource farms, and larger farms experiencing significant distress due to natural disasters and economic emergencies. Those farmers do not have access to commercial credit; thus, they are considered candidates for public credit programs. In turn, the federal government responded with a lengthy series of legislative enactments and agency developments that included the creation of the Farmers Home Administration in 1946.[2]

Although special credit programs and other aids to low-income farmers in certain emergency situations had been available since before World War I, the groundwork for institutionalizing federal credit programs for agriculture occurred during the depression years of the 1930s. The very titles (rehabilitation, resettlement, security, subsistence) of the agencies and legislation showed the needs of the period and the evolution of the programs as economic conditions changed (see Barry and Boehlje [1986] for further details). Under the Farmers Home Administration Act of 1946, almost all the direct lending to low-income farmers that had been carried out for several years under a variety of programs and agencies was placed under the newly formed FmHA.

Besides the direct financing of farmers' operating and ownership needs, through insured loans and later guarantees of farm loans made by commercial lenders, FmHA also acquired expanded authorities for making emergency loans. One type of emergency was a major disaster affecting selected areas or regions and usually attributed to natural causes (droughts, hurricanes, tornadoes, floods, fires, and weather-induced insect infestations and diseases). The other type of emergency was the severe stress imposed by economic and credit conditions. The concept of an economic emergency had surfaced briefly in the early 1950s and then again in the 1970s, first with passage of

2. USDA restructuring in late 1994 has resulted in the termination of the Farmers Home Administration and the transfer of its farm loan programs to the new Consolidated Farm Service Agency. In this section, however, the name Farmers Home Administration will continue to be used for familiarity and in reference to past FmHA programs.

the Emergency Livestock Credit Act of 1974 (to guarantee loans made by commercial lenders to financially distressed livestock and poultry producers—terminated in 1979) and then the Agricultural Credit Act of 1978. The 1978 act introduced the concept of an economic emergency, defined as a condition resulting from a general tightening of agricultural credit or an unfavorable relationship between production costs and prices received for agricultural commodities, which resulted in widespread need among farmers for temporary credit. The status of economic emergency was intended to be temporary, but it was extended several times during the financial stresses of the 1980s.

Beginning in 1949, FmHA received legislative authorization for an enlarged scope of financing activities that continued to increase in the following three decades. These authorizations extended the agency's credit programs to the nonfarm activities of rural residents and rural communities. Housing loans for farmers, first authorized in 1949, were later available to nonfarm residents. Financing for water facilities was extended to nonfarm rural customers and then to rural communities. Financing of waste disposal systems was added as well. In the 1960s, FmHA also undertook the financing of rural development programs, which included economic opportunity loans to low-income rural people for farm and nonfarm enterprises and aiding communities to attract new industry through community facility loans.

Following the Rural Development Act of 1972, FmHA's involvement in rural development expanded even more (Herr and LaDue 1981). This act authorized FmHA to guarantee loans made by commercial lenders for farming, housing, and rural business and industries and greatly expanded the loan limits and availability of water and waste disposal loans (Meekhof 1984). FmHA could also make loans for rural community facilities such as fire departments, hospitals, nursing homes, and public recreation centers. In some cases, grants could be made for water, waste disposal, and certain other programs, including the improvement of rural industrial sites.

As a brief historical footnote, for a time in the 1970s, the Small Business Administration (SBA) also received authority to finance eligible agricultural producers—a political development that allowed two federal agencies to provide duplicate credit services and essentially to compete with each other in offering public credit programs for agriculture. Beginning in 1980, regulations were established to coordinate better the agricultural lending activities of the SBA and FmHA, and soon thereafter SBA's involvement in agriculture was largely terminated. Nonetheless, SBA's loan volume to farmers was considerable, reaching an estimated $3 billion of debt in 1981, with much of this debt still outstanding later in the 1980s.

The characteristics of the farmers receiving public credit were also changing. According to a USDA study of FmHA borrowers in 1979, the traditional farm operating and farm ownership loans were directed principally to young farmers and to those with small net worths and low incomes (Lee and Gabriel 1980). In contrast, much of the funding for FmHA's economic emergency program went to farmers with relatively large net worths, high debt loads, and strong income potential (but low current incomes). Moreover, the emergency concept also resulted in FmHA refinancing a considerable number of problem loans for commercial banks, the FCS, and other lenders that alleviated, at least for a time, deterioration in these institutions' financial conditions.

As FmHA (and SBA) loan volume grew substantially in the late 1970s and early 1980s, the personnel capacity of the agencies was stretched to a limit. Delays in handling loan applications occurred, and the agencies came under criticism for their lending practices. Administrative resources and practices eventually improved, but it was an uphill struggle.

Downsizing versus Responding to Financial Stress

At the beginning of the 1980s, there were a clear understanding and widespread agreement that direct public lending through FmHA to agriculture had become excessive and needed curtailment to restore the agency's last resort role (Barry 1985). Curtailment at that time seemed feasible for several reasons: (1) high taxpayer costs of public credit and other farm programs; (2) stronger financial performance anticipated for the farm sector; (3) slower growth in farm debt; (4) expanded use of the revised Federal Crop Insurance Program for disaster protection; (5) more effective use of loan guarantees in public credit programs; and (6) more effective risk management by agricultural lenders.

High levels of financial stress in the 1980s, political pressures involving farm credit, and ineffectiveness of the crop insurance program, however, thwarted this redefinition of FmHA lending. Instead of downsizing, FmHA became a significant provider of additional financial assistance to the agricultural sector and even now continues to carry substantial adversely classified credit in its loan portfolio. It is not surprising, however, that public credit would increase in stress times and that the lingering effects of the financial stresses of the 1980s would weigh more heavily on the lender of last resort, in contrast with the quicker recoveries of agricultural banks and the Farm Credit System. In addition, as indicated above, the significant amount of federal assistance channeled through FmHA in the 1980s probably reduced the adversity the FCS and agricultural banks experienced. In the absence of FmHA, the need for financial assistance by the FCS would probably have been much greater than the $1.26 billion actually used.

The loan loss experiences of FmHA relative to other lenders are shown in table 6–1. FmHA loan losses totaled $16.19 billion over the 1986–1993 period, compared with losses of $3.41 billion for the FCS, $2.18 billion for commercial

TABLE 6–1
FARM LOAN LOSSES (NET CHARGE-OFFS), BY LENDER, 1984–1993

Year	Commercial Banks[a]		Farm Credit System[b]		Farmers Home Administration[c]		Exhibit: Life Insurance Company Foreclosures[d]	
	$	%	$	%	$	%	$	%
1984	900	(2.3)	428	(0.5)	128	(0.5)	289	(2.5)
1985	1,300	(3.3)	1,105	(1.6)	257	(0.9)	530	(4.8)
1986	1,195	(3.4)	1,321	(2.3)	434	(1.5)	827	(7.9)
1987	503	(1.6)	488	(0.9)	1,199	(4.3)	692	(7.5)
1988	128	(0.4)	413	(0.8)	2,113	(8.4)	364	(4.0)
1989	91	(0.3)	(5)	(0.0)[e]	3,297	(12.4)	204	(2.3)
1990	51	(0.2)	21	(0.04)	3,199	(13.5)	85	(0.9)
1991	105	(0.3)	47	(0.09)	2,289	(10.4)	95	(1.0)
1992	82	(0.2)	19	(0.04)	1,887	(9.1)	148	(1.8)
1993[f]	23	(0.0)[g]	4	(0.0)[e]	1,768	(9.4)	78	(0.9)

NA = Not available.
NOTE: Loan loss data rounded to nearest million dollars.
a. Calendar year data for non–real estate loans.
b. Calendar year data.
c. Fiscal year data beginning October 1. Includes data on the insured (direct) and guaranteed farm loan programs.
d. Loan charge-off data are not available for life insurance companies.
e. A gain of less than 0.01 percent.
f. Commercial bank data through June 30, 1992, and Farm Credit System and life insurance company data through September 30, 1992.
g. Less than 0.05 percent.
SOURCES: American Council of Life Insurance, Board of Governors of the Federal Reserve System, the Farm Credit Council, and Farmers Home Administration.

banks, and $2.49 billion for life insurance companies. One significant difference in historical loss accounting among these lenders is that FmHA included accrued, unpaid interest in its loss totals. In contrast, once banks and the FCS place loans in a nonaccrual status, nonaccrued interest is not counted as a loss.

Current FmHA Developments

By 1994, FmHA had achieved two major accomplishments. First, the amount of credit channeled through its

61

farm programs was substantially reduced. As shown in tables 6–2 and 6–3, new loan obligations declined from $6.28 billion in 1980 to $5.92 billion in 1985, $2.17 billion in 1990, and $2.13 billion in 1993. Similarly, farm debt outstanding on direct loans from FmHA declined from $24.54 billion in 1980 to $12.21 billion in 1993, although a major part of this reduction is attributed to loan losses.

Second, the form of FmHA lending swung significantly toward guaranteed loans rather than direct loans. Considering the farm ownership and operating loans only, new obligations were 97.1 percent direct loans and 2.9 percent guaranteed loans in 1980, 78.4 percent direct and 21.6 percent guaranteed in 1985, and 29.5 percent direct and 70.5 percent guaranteed in 1993. The shift toward guaranteed loans is consistent with the Office of Management and Budget mandate of 1984 and legislative enactments in 1985. The shift reflects the lower subsidy cost and other merits of guaranteed loans discussed earlier. FmHA personnel report, for example, that subsidy costs in mid-1994 were 8.67 percent for direct loans, 0.21 percent for guaranteed loans, and 5.22 percent for loans to limited-resource farms (a broader range of subsidy costs for public credit programs is found in tables 3–4, 5–1, and 5–2). The agency is sensitive to these subsidy costs and to the concept of cost control along with an operating goal of serving as many farm borrowers as possible given the size of the annual allocations.

Graduation Attributes of FmHA Programs

A continuing issue in FmHA farm lending programs has been the length of participation by individual borrowers and the prospects for "graduating" these borrowers to commercial credit. FmHA regulations define graduation as "the payment in full of an FmHA loan before maturity by refinancing through other credit sources." Graduation is the ultimate objective of a public credit program in which eligible borrowers are presumed to have potential for future viability and development of creditworthiness sufficient to qualify for commercial financing. Since 1989, FmHA has used a credit-scoring model for its direct loan borrowers, taking account of five factors—FmHA security margin in collateral pledged as loan security, the borrower's debt-to-asset ratio, the current ratio, return on assets, and repayment ability—to evaluate the quality of its direct loan portfolio. The model results place borrowers in one of five credit classes: commercial, standard, substandard, doubtful, and loss.

The commercial class is defined to include borrowers who would appear acceptable for financing by commercial lenders. FmHA personnel in county or local offices assess graduation potential by following several steps: they monitor fund availability and credit conditions in their local credit markets, they build familiarity and rapport with local agricultural lenders (primarily commercial banks and FCS lending associations), and they periodically send loan documentation of commercial-classed borrowers to local lenders for possible acceptance as a loan customer. If the commercial lenders do not accept the applicant, the applicant continues as an FmHA borrower.

The recently adopted credit classification system improves on the old FmHA guidelines in which a specified percentage (that is, 10 percent) of the local office's borrowers were reviewed for possible submission to local commercial lenders. Nonetheless, under both the old and the new approaches, the graduation decision still rests with the acceptance or rejection decisions of commercial lenders, who evaluate the potential profitability of the FmHA borrower as a new loan applicant. The process is also vulnerable to moral hazards (LaDue 1990) by the farm borrower who may prefer to remain eligible for the more favorable, subsidized credit terms of FmHA. The lender

TABLE 6-2
ALLOCATIONS OF THE FARMERS HOME ADMINISTRATION LOAN PROGRAMS, FISCAL YEARS 1980–1994
(thousands of dollars)

Year	Direct Operating Loan	Guaranteed Operating Loan	Direct Farm Ownership Loan	Guaranteed Farm Ownership Loan
1980	850,000	25,000	820,000	50,000
1981	850,000	25,000	870,000	50,000
1982	1,325,000	50,000	700,000	125,000
1983	1,460,000	50,000	700,000	75,000
1984	1,810,000	100,000	625,000	50,000
1985	1,920,000	650,000	650,000	50,000
1986	1,740,000	1,660,900	388,820	249,250
1987	1,425,000	2,170,000	75,000	325,000
1988	900,000	2,400,000	115,000	390,000
1989	900,000	2,300,000	95,000	474,000
1990	932,500	2,565,985	80,000	459,279
1991	493,300	2,854,700	57,200	783,300
1992	850,000	1,982,140	66,750	488,750
1993	825,000	1,738,354	66,750	488,750
1994	700,000	2,050,000	78,081	556,543

SOURCE: Farmers Home Administration, U.S. Department of Agriculture, 1994.

64

TABLE 6–3
Obligations of the Farmers Home Administration Loan Programs, Fiscal Years 1980–1993
(thousands of dollars)

Year	Direct Operating Loan	Guaranteed Operating Loan	Direct Farm Operating Loan	Guaranteed Farm Operating Loan	Emergency Disaster	Direct Economic Emergency	Guaranteed Economic Emergency
1980	849,999	24,830	926,198	27,854	2,266,890	2,076,612	108,870
1981	822,614	24,989	795,353	17,932	5,112,290	1,160,672	84,681
1982	1,203,679	47,329	657,747	3,856	2,173,412	0	0
1983	1,684,999	50,547	729,546	20,032	565,937	0	0
1984	1,959,709	111,444	659,191	41,504	1,051,627	309,388	289,944
1985	3,599,999	1,106,849	651,870	67,926	490,876		
1986	2,203,178	1,367,286	371,388	192,018	217,774		
1987	1,298,262	1,240,738	74,998	324,419	113,612		
1988	899,500	892,578	114,978	362,086	29,890		
1989	886,143	886,143	94,932	305,062	73,492		
1990	733,291	908,747	79,983	348,719	101,509		
1991	489,909	1,035,282	57,139	365,511	81,402		
1992	570,736	1,107,914	66,658	452,391	74,883		
1993	545,173	1,013,340	66,813	448,953	58,607		

Note: Direct and guaranteed economic emergency loans were terminated after 1984.
Source: Farmers Home Administration, U.S. Department of Agriculture, 1994.

too may exhibit moral-hazard behavior by preferring to reject a borderline borrower, knowing that the borrower may continue as an FmHA client.

A new initiative being developed by FmHA is the development of a comprehensive business plan, eventually for all of its borrowers but initially for new and high-risk borrowers. The documentation contained in the business plan should help the graduation process.

Several other options could be considered to facilitate graduation. FmHA could adopt provisions that would, after some period, change graduation from a discretionary decision to a mandatory discontinuation of financing, either by graduation or departure from farming. A maximum length could be placed on an individual borrower's participation in the farmer program, as with the new young farmer program, in which ten- and fifteen-year limits are placed on the participation in the direct and guaranteed-loan programs, respectively. This maximum length, however, should not become the norm; instead, to the extent possible, earlier graduation should be the emphasis. Using the credit-scoring models of local commercial lenders rather than FmHA's alone could be part of this process.

Providing rewards and incentives for borrower progress and graduation could also receive greater consideration. The borrower's financing costs, for example, could move downward with progress, with a partial rebate of interest payments when early graduation occurs or according to other terms agreed on by FmHA and the borrower. FmHA could also compensate the private lender for accepting the borrower by paying part of the borrower's interest payments for a stipulated period—similar to the interest buy-down programs of the 1980s. Or local FmHA personnel could provide rewards for the speed, efficiency, permanency, and other attributes with which FmHA borrowers move toward graduation into private sector lending. Any personnel incentives associated with program magnitude (number of borrowers served, loan volume, staff size) could be deemphasized. Of course, the disincentive here is that complete success in graduating all FmHA borrowers would end the need for FmHA loan personnel.

The shift to FmHA loan guarantees facilitates graduation by providing a bridge between the direct loan status and complete graduation. With a guarantee, the borrower becomes a direct customer of the commercial lender, and the goal is then to provide a bridging of credit risks that eventually qualifies the borrower for an unfettered credit relationship with the commercial lender. Periodic expirations of guarantees or potential refinancing to private status may flow more naturally from the guarantee status than from the direct loan involvement of FmHA.

Finally, eligibility for FmHA credit might be based on a first refusal by the guaranteed loans from credit programs provided by individual states. Many states have such programs, although their small size and limited personnel would hinder this approach. Perhaps a federal-state partnership program could be developed to shift more of the funding and administration to the states.

USDA Restructuring

The USDA restructuring proposed by Secretary of Agriculture Mike Espy and signed into law in 1994 also has important implications for public credit programs for agriculture. Under restructuring, FmHA has been dissolved as a USDA agency with the farmer programs transferred to a Consolidated Farm Service Agency (along with the Agricultural Stabilization and Conservation Service and the Federal Crop Insurance Corporation), and the rural housing and business development loans are being transferred to a rural development agency. In principle, the service capacity to farmers would be enhanced by consolidating those programs into one local office, and the types of farm loans would remain the same. Moreover, the number of contact points of FmHA programs with rural communities has increased as a result of the restructuring.

Considerable uncertainty remains, however, about how much those in the local offices know about credit; about the role of county committees composed of farmers, which have currently been disbanded; and about the adjustment costs of consolidations of dissimilar data, computer, communication, other information, and employment systems across the consolidated agencies. These details can disrupt the transition process and detract from the quality of credit decisions and loan monitoring over the long term.

Management, Control, and Risk Exposure

In recent years, FmHA has come under substantial scrutiny and criticism regarding the management, control, and risk exposure of its loan programs. FmHA faces a nearly chronic dilemma of balancing a lender-of-last-resort and an emergency credit role against the need to operate as an efficient, responsive agency that maintains effective programmatic oversight and control and that ensures full compliance of operations by its lending personnel with guidelines and directed procedures.

On the one hand, FmHA has borne the brunt of the public credit syndrome in which credit programs are popular responses to all sorts of financial adversity. The financial horror stories of the 1980s are well known, and it simply became unacceptable (illegal for a time because of foreclosure moratoriums) for FmHA to put distressed farmers out of business. The agency was asked to take on a massive relief mission without commensurate growth in administrative resources. It may have functioned as best it could under the circumstances, especially in light of the rather weak and ambiguous signals received from Congress about the agency's mission and goals. While FmHA is intended to function as a lender of last resort providing a temporary source of credit, it has no clear statutory guidelines in defining the conditions under which the agency can resolve its relationship with many types of borrowers.

On the other hand, FmHA has experienced substantial criticism over the years regarding the management, control, and risk exposure of its lending programs. A series of reports of the General Accounting Office (October 1991; April 1992) have cited the agency's inability to develop an effective information system (despite substantial financial resources devoted to the effort) and its longstanding planning and oversight problems and have raised serious questions about the agency's planning process. The criticisms extend to the management and control of the agency's loan programs. In an April 1992 report, GAO (p. 5) states:

> The multibillion-dollar federal investment in farmer loan programs is not being adequately protected. In the direct loan program, field lending officials have not complied with agency loan-making and loan-servicing standards established to safeguard federal financial interests. In addition, FmHA's loan-making and loan servicing policies—designed, in large part, to help farmers stay in farming—have increased the government's losses. By allowing delinquent borrowers to obtain additional credit, FmHA has reinforced its lending to poor credit risks, and by providing debt relief to borrowers who have defaulted on their loans, it has created incentives for farmers to avoid repaying their debts.
>
> In the guaranteed loan program, ineffective implementation of agency standards and imprudent policies have also jeopardized the fed-

eral investment. FmHA lending officials have approved guidelines without obtaining proof of borrowers' credit worthiness and have not adequately monitored commercial lenders' servicing of guaranteed loans. Additionally, policies permitting commercial lenders to refinance existing farm debt and obtain maximum-rate guarantees for most loans, regardless of risk, have encouraged lenders to shift their high-risk farm debt to the government.

The report further states:

> By almost any measure, FmHA's loan programs have become good examples of how programs should not be implemented and managed. Because legislation has not established clear priorities for FmHA's mission, the agency has tried simultaneously to meet conflicting objectives—to be fiscally prudent and to provide high-risk borrowers with temporary credit to keep them in farming until they secure commercial credit. Arguably, FmHA has not achieved either objective. Its shaky loan portfolio does not reflect the operations of a prudent lender. Furthermore, as an assistance agency, FmHA has had little success in graduating borrowers to commercial sources of credit, as was originally anticipated. Ironically, some of FmHA's clients are financially weaker after FmHA's help than before.

Of course, many positive contributions and success stories for FmHA borrowers could be cited as well, but the concerns have become large issues.

These criticisms go beyond the breakdown of a principal-agent relationship through asymmetric information problems and misaligned incentives. They exemplify the classic case of a credit program asked to carry relief and welfare functions well beyond its proper scope and mission. The effective workings of a credit market are undermined by asking FmHA to bear too much ill-defined responsibility aimed in part at keeping marginal farmers in business.

Moreover, the agency does not have sufficient resources or types of loan personnel to function consistently as a supplement to commercial lenders—that is, as a temporary repository of soon-to-be-creditworthy borrowers. Instead, a credit agency is serving a welfare purpose.

7
The Farm Credit System

The Farm Credit System is a system of federally chartered, privately owned banks, lending associations, and service units organized as cooperatives with the purpose of providing credit and related services to agricultural producers, rural homeowners, and agricultural cooperatives in the United States.

Mission and Scope

By virtue of the cooperative organization, the agricultural borrowers become the system's owners and holders of its equity capital, and they are represented by elected boards of directors. The institutions of the FCS are regulated and examined by the Farm Credit Administration, an independent agency in the executive branch of the U.S. government.

Congressional authority for the FCS, contained in the Farm Credit Act of 1971, as amended, specifies that the system work to improve the income and well-being of American farmers and ranchers by furnishing sound, adequate, and constructive credit and closely related services to creditworthy borrowers and that it provide these services through favorable and unfavorable times. The mandate to be a reliable agricultural lender creates the need for a reliable source of funds and a heavy emphasis on risk management through the system. The farm credit banks acquire most of their loanable funds by issuing debt securities (federal farm credit banks consolidated systemwide bonds, medium-term notes, and discount notes) in the national financial markets. The debt securities are treated as "government agency securities" in the financial market, even though it is explicitly stated that they are not guaranteed by the federal government against default. Agency status results from a set of regulatory exemptions and preferences of these securities, as they are traded in the financial markets, and the perception of implied government backing if the system experiences severe financial difficulty. Agency status is a significant factor in the ability of the farm credit banks to market large volumes of securities at relatively favorable costs.

As shown in an earlier section of this volume, the institutions of the FCS have long played a major role in financing U.S. agriculture. Total FCS loan volume to agricultural producers reached a peak of $64.5 billion in 1982, or 34 percent of total farm debt, before declining to a low of $40.4 billion (25 percent of farm debt) in 1991. About three-fourths of the FCS debt is real estate debt, indicating the strong importance of the FCS in meeting the real estate financing needs of agricultural producers. In addition, loans outstanding from the banks for cooperatives totaled about $14.8 billion at year end 1993.

Performance and Restructuring

The FCS developed from 1916 to the mid-1930s as a specialized lender to agriculture to provide appropriate real estate financing terms for farmers and an alternative source of financing for agricultural production (and agricultural cooperatives) to the then unstable system of commercial banking in rural America. In accomplishing its objectives, the FCS as a whole has experienced periodic changes since the 1930s in the structure of its lending institutions and regulatory agency to revitalize financial performance, streamline operations, restructure business units, improve efficiency, manage risks, and compete effectively in the future. The late 1980s and early 1990s became a period of significant, accelerated structural change for the FCS because of the combination of significant financial stress in agriculture, financial problems affecting FCS institutions, and greater competition among financial institutions induced, in part, by regulatory change (Barry and Lee 1983).

The 1980s saw substantial downsizing of the FCS loan portfolio, loan losses totaling about $3.41 billion, and additional economic losses of a greater amount because of inadequate loan-pricing policies in a volatile interest rate environment. Loans to the system's borrowers were priced with variable interest rates adjusted periodically in response to changes in an institution's average cost of funds. Thus, FCS loan rates lagged behind market rates, remaining lower on average as market rates increased, and then rising above falling market rates and declining only slowly thereafter. During the mid-1980s, as market interest rates began to decline from the record high 15 to 20 percent range, FCS rates, which had never reached this range, continued to increase and, though eventually declining, stayed above rates from competing banks, trade firms, and other lenders for some time. The combination of good borrowers' leaving the FCS to refinance with other lower-rate lenders and rollovers of maturing bonds at high interest rates locked the system into high funding costs and adversely affected the financial performance of many FCS institutions.

In addition, while all the farm credit banks were responsible for the system's debt security sales to investors, some of the more financially healthy banks were reluctant to provide assistance to their troubled sister institutions. In some cases, lawsuits were filed to prevent intrasystem transfers of financial capital. As a result of these difficulties, the Agricultural Credit Act of 1987 created mechanisms to provide financial assistance from the

U.S. government to FCS institutions that needed outside help to redeem borrower stock at par, to pay off debt securities as they matured, and to continue lending at competitive rates. By 1993, approximately $1.26 billion of federal financial assistance had been provided to the FCS institutions. The subsequent rapid financial recovery of institutions in the system is supporting an early payback of this assistance.

Substantial restructuring of the FCS has occurred in the past ten years. Some was motivated by the system's own self-study (Project 1995), but most is attributable to the financial stresses of the 1980s, the need for greater operating efficiencies, and the resulting Farm Credit Act Amendments of 1985 and 1986 and the Agricultural Credit Act of 1987. Because of consolidations and restructuring, lending associations are now much fewer in number, larger in size, and more diverse in legal structures. The number of lending associations, for example, declined from 915 in 1980 to 240 by 1994. One farm credit district now has only a single, districtwide agricultural credit association, and several others have large multistate associations. Other districts include varying combinations of agricultural credit associations, federal land credit associations, production credit associations, and federal land bank associations. In many cases, the real estate and non–real estate lending associations have common territories and management and in the case of agricultural credit associations are characterized by a fully consolidated institution.

In all the farm credit districts, the 1987 act required the old federal land banks and federal intermediate credit banks to merge to form farm credit banks. Some interdistrict mergers of farm credit banks have occurred as well, so that in 1994 there were eight farm credit districts, with additional mergers likely in the future. The 1987 act also required the twelve district banks for cooperatives to vote whether to merge with the central bank for cooperatives or to remain separate. Ten of the banks for cooperatives elected to consolidate, forming a new national bank for cooperatives called CoBank. Two banks for cooperatives (Springfield and St. Paul) remained separate, although the Springfield Farm Credit Bank and the Springfield Bank for Cooperatives subsequently merged with CoBank. The two remaining banks for cooperatives now have national charters and may operate on a national basis with competing territories, if they so choose.

The legislation and related experiences of the 1980s brought several other major changes to the FCS. The Farm Credit Administration became an arm's-length regulator with more stringent powers for regulating and evaluating the safety and soundness of the FCS institutions. Thus, the FCS and FCA are now structured like federal regulators for other types of financial institutions.

Equity capital of the various banks and associations is obtained by sale of stock to borrowers and by retained earnings. Before the 1987 act, agricultural borrowers were required to purchase stock in their local association at rates that fell between 5 percent and 10 percent of their loan amount (the funds for the purchases were held back as part of the loan). The value of the stock was fixed and largely considered risk free (that is, protected). After 1987, the required minimum investment of each borrower was substantially reduced—to $1,000 or 2 percent of the loan amount, whichever is less. Virtually all the equity capital of the lending associations is now considered at risk and largely generated by retained earnings. Many associations, however, maintain borrower stock requirements at levels above the minimum. The 1987 act also phased in minimum capital requirements for FCS institutions based on an equity-to-asset ratio of 7 percent or greater. Currently, all FCS institutions have capital ratios well above this minimum.

The 1987 act also authorized the establishment of the Farm Credit System Insurance Corporation (FCSIC) to provide a safety reserve for investors in farm credit securities in case of financial distress experienced by the FCS institutions. The FCSIC will provide an alternative and prior risk control mechanism to the joint and several liability concept of the farm credit banks that, itself, was plagued by problems of peer monitoring and control during the 1980s.[1] The insurance program will probably add to investors' perceptions of safety and soundness in the system and reduce the likelihood of demands for public financial assistance for the FCS in the future.

Besides the creation of the FCSIC, the farm credit banks and the banks for cooperatives also undertook some collective, self-initiated actions to build financial control, discipline, safety, and soundness across the respective banks. The actions include the Contractual Interbank Performance Agreement, established in 1991, and the recently approved Market Access Agreement. Under the performance agreement, the insured banks delegated to the Federal Farm Credit Banks Funding Corporation the power to establish accounting and financial standards (in addition to those of the FCA) for the banks and to impose additional financial discipline on the banks. Each bank's financial performance is scored monthly with various penalties imposed on low-scoring banks. The Market Access Agreement provides a mechanism for limiting a bank's participation in the sale of systemwide, consolidated securities if bank performance deteriorates significantly. While the effectiveness of these risk-control mechanisms remains untested, they represent significant steps toward effective, internal surveillance and control by the responsible institutions.

Despite the significant changes, the basic structure of the FCS still reflects a tiered approach to ownership, management, and funds distribution. The agricultural borrower acquires an ownership interest in the local association and participates in the selection of management by voting for the association's board of directors. In turn,

---

1. "Joint and several" means that all of the farm credit banks and banks for cooperatives are responsible for meeting obligations on the banks' securities sold to investors in the financial market.

the association either obtains its funds from, or serves as a lending agent of, the district farm credit bank. Along with this funding arrangement, the local association acquires an ownership interest in the district farm credit bank and participates in the selection of its management by voting for the bank's board of directors. In the case of the banks for cooperatives, the link between the borrowing cooperative and the bank is a direct one; no lending associations are found in the banks for cooperatives system, although a recent merger among banks has made CoBank a provider of loan funds to a number of lending associations in the northeastern United States. Finally, the farm credit banks join together with the banks for cooperatives to issue the systemwide, consolidated debt securities in the financial market, with joint and several liability among the banks (but not the lending associations) for these debt securities.

Risk and Risk Management

The mission to provide specialized credit services to an inherently unstable agricultural industry has placed a high priority on the monitoring and management of risks experienced by the FCS institutions. This priority was further heightened by the severe financial stresses of the 1980s. Especially important have been the effects of credit risk, interest rate risk, and liquidity risk.

Credit risk is the potential delinquency or default on repayment by borrowers and the potential insufficiency of collateral pledged as security to cover outstanding loan balances. Many of the activities and responsibilities of system personnel are directed toward safe and sound lending and the management of credit risks, as are many of the changes resulting from the 1987 act.

The effects of credit risk can be transferred throughout the system, ultimately to jeopardize payments of principal and interest to the investors in farm credit securities and to trigger requests for federal assistance. Currently, however, an extensive set of protection mechanisms is in place to deal with credit risk. These mechanisms can be ordered as follows, starting with the agricultural borrowers and ending with the FCS banks' joint and several liability for the farm credit securities:

1. comprehensive evaluations of individual borrower's creditworthiness to determine:
   - adequacy of the borrower's earnings and willingness to make payments, subject to the borrower's skills in risk management
   - adequacy of collateral and ability to realize proceeds from the sale of secured assets
2. conservatism of lending terms as indicated by higher downpayments from borrowers, shorter maturities on longer-term loans, more rapid repayment plans (for example, equal principal payments versus equally amortized payments), risk-adjusted interest rates, and others
3. bank reviews of association loan portfolios
4. Farm Credit Administration examinations of banks and associations
5. geographic restructuring and mergers that have increased institutional risk-bearing capacity
6. financial structure and reserves of banks and associations
   - asset liquidity structure: total loans to total assets
   - institutional leverage positions: equity capital to total assets
7. risk-sharing agreements among selected FCS institutions
8. self-discipline through the Contractual Interbank Performance Agreement and the Market Access Agreement
9. adequacy of the FCSIC insurance fund
10. joint and several liability of the system banks

The availability and effective use of these protection mechanisms forestall excessive credit risk and respond to the normal range of credit risk—that is, through the normal swings of the agricultural business cycle. Improvements in risk management skills and financial documentation by many agricultural borrowers have also helped FCS to manage credit risks.

Interest rate risk arises from the volatility of market interest rates and their effects on the net interest margins and market values of equity of system institutions. Interest rate risks are not a major problem if the rate-sensitive assets and liabilities used to fund those assets are well matched in maturity or in other more technical characteristics (called "duration" in the finance literature). If assets and liabilities are well matched, then both the beneficial and the adverse effects of unanticipated rate changes are offsetting, and institutional financial performance is stabilized. If they are not well matched, then wide swings in net interest margins and institutional net worth may occur.

Matching of assets and liabilities may be accomplished in several ways—through equal holdings of similar assets and liabilities, floating or administered pricing policies on loans and securities, use of financial futures, and use of interest rate swaps and other derivatives. In the 1980s, the FCS was victimized by high interest rate risk due to significant mismatching of assets and liabilities at that time and to the average cost pricing policy described earlier. Since that time, however, system institutions have substantially upgraded their skills in managing interest rate risk (called asset-liability management) so that its potential effects are substantially reduced. The farm credit banks, with the advice and counsel of their funding entity (the Federal Farm Credit Banks Funding Corporation), are now responsible for asset-liability (A-L) management on behalf of the lending associations in their respective districts. The banks' large sizes enable them to acquire the necessary personnel and other resources to carry on effective A-L management. These attributes do not remove interest rate risk entirely, but they do reflect major gains in its management.

Liquidity risk is the inability to generate funds to meet financial obligations, including possible loss of access to specified sources of funds. Part of liquidity risk is related to interest rate risk, because the market value of investment holdings by FCS institutions may have declined through increases in market interest rates at the very time that cash may be needed to meet principal and interest obligations on farm credit securities. Another part of liquidity risk for the FCS reflects the potential loss of market access because of changes in the agency status of farm credit securities, including investors' perceptions of a loss of implied government support for these securities. Agency status issues are considered in the following section.

In general, then, as a result of the stresses of the 1980s, the risk management activities of the Farm Credit System and the FCA have been substantially strengthened, especially in dealing with the continuing risks of agricultural lending and volatility in the financial markets. The catastrophic nature of risks facing the FCS, however, is difficult to insure against. The incidence of such severe risks, which may reflect the combined effects of credit risk, interest rate risk, liquidity risk, and other sources of risk, tends to be concentrated in relatively short periods of time. The financial adversities of the 1980s, for example, were the first major loss experience since the 1930s, indicating a loss concentration covering a four- to five-year subset out of approximately fifty years. It remains unclear whether risk protections outlined in this section can prevent a similar recurrence during the next fifty years.

Agency Status Issues

While the FCS is considered privately owned and operates much like other commercial finance institutions, its major source of funds is from the sale of farm credit securities that are treated as "government agency securities" in the financial markets (Barry 1984; Lins and Barry 1984). Agency status is accorded these securities even though it is explicitly stated in the statute that they are not guaranteed against default by the U.S. government. Agency status results from regulatory exemptions and trading preferences of these securities and the perception of implied government backing if the system experiences severe financial difficulties.

Also important to accessibility to the financial markets are the system's credit history, financial structure, and an efficient, well-managed distribution system for the securities. The latter are considered necessary but not sufficient conditions for market access. The elements of agency status are the sufficient conditions; they are considered a significant factor in the ability of the FCS to market large volumes of securities at relatively favorable costs. Agency securities generally trade at yields that fall between the yields on U.S. Treasury securities and the yields on prime corporate bonds of comparable maturities—for example, usually a ten to fifty basis point spread over U.S. Treasury securities.

Agency status of the Farm Credit System and other government-sponsored enterprises has been closely scrutinized over the past fifteen years, with a view toward eventual and complete privatization of these institutions. This scrutiny reflects questions about the general role of and specific need for government-sponsored enterprises, the missions of the respective institutions, and the implied contingent liability of the U.S. government in upholding the financial obligations of those enterprises. As indicated earlier, agency status for the FCS may ensure reliable access for funding in return for the system's highly concentrated, single industry (agriculture) loan portfolio. Moreover, continuation of agency status has seemed appropriate during the system's recent financial recovery.

Thus, it is likely that agency status will continue to invite policy debate in the future.

Expanded Authorizations

Confining the mission of the FCS primarily to financing agricultural production and agricultural cooperatives in the United States results in a significant concentration of risk in the system's loan portfolio. Consideration is occasionally given to broadening the range of eligible borrowers to increase the system's risk-carrying capacity and more effectively meet the financing needs of rural America. Modest increases in FCS lending authority have occurred over time. The Farm Credit Amendments Act of 1980 allowed system institutions to finance the on-farm marketing and processing or farm-related business activities of eligible borrowers, with 1990 legislation allowing additional financing of agribusiness activities based on throughput requirements for previously eligible agricultural borrowers. Other broadening of lending authority has included rural housing loans, international lending by the banks for cooperatives, loans to aquatic and timber producers, and the ability of the banks for cooperatives to provide credit enhancements for some types of rural development lending.

In 1994, two pieces of legislation were introduced that would further broaden the lending authority of the FCS institutions. The provisions of the Rural Credit and Development Act of 1994 (called the Clayton Bill) include: (1) broadening authority to lend to rural and agricultural businesses that provide related goods and services to farmers and ranchers; (2) permission for FCS institutions to purchase such loans originated by others; (3) broadening the authority to lend to businesses serving cooperatives that are, in turn, eligible for FCS financing; (4) authority for FCS institutions to lend to rural communities for facility projects; (5) authority for FCS lending to rural utilities; and (6) expanded authority to provide rural home mortgage credit by increasing the present community population limitation of 2,500 to 20,000 and increasing the portfolio limitation from 15 percent to 20 percent.

The second piece of legislation (the CoBank Bill), which was enacted into law, removes or reduces a number of restrictions on the banks for cooperatives with respect to joint venture financing, import and export financing, and loan participation authority. Major changes include: (1) financing eligible cooperatives rather than just stockholders; (2) financing all eligible agricultural exports, farm supplies, or aquatic products, including freight rather than only exports of stockholders with specified minimum ownership interests; and (3) broadened participation in any prescribed multilender transaction rather than those in which loan participations do not exceed 10 percent of bank capital and 50 percent of loan principal.

The long-run goals of these proposals are to enhance the risk-carrying capacity of system institutions and to provide a competitive source of credit from the national financial markets for rural borrowers, communities, and cooperatives whose financing needs are otherwise unmet or inequitably served. The extent of risk reduction in the Clayton Bill is difficult to project because the financial health of agribusinesses and many rural communities is often strongly correlated with that of agricultural production. Moreover, credit risks might actually increase in the near term from a lack of familiarity by FCS institutions with new types of borrowers. In addition, opponents to these proposals have argued that existing credit sources are available to meet these financing needs adequately, especially from commercial banks, and that expanding lending authority might detract from the FCS financing of its traditional agricultural clientele.

Judging from past experience, new lending authorities for FCS institutions will likely continue to occur, although incrementally. Thus, some broadening of future lending authority is logical to expect. Often, however, FCS institutions must reach a compromise with competitors in the political process to gain broader authority. Such a compromise might involve acceptance of more stringent controls and capital requirements, agreement to offer a publicly mandated credit program, or concession of part or all of agency status.

In contrast with that of other countries, the U.S. Farm Credit System is more narrowly defined. Some of the world's largest agricultural lenders (for example, Credit Agricole and Rabbobank) were originally organized as local cooperative banks that now have international deposit-gathering and lending authority. The home environment of these international banks, however, did not typically include a decentralized commercial banking system inclined to finance agriculture or serve rural residents, compared with the community-oriented banks in the United States. Thus, more broadly based rural credit institutions were needed in these countries. The need for such breadth has been much less in the United States.

Outlook

The Farm Credit System's major contribution has been the provision of reliable, long-term credit to agricultural borrowers to finance the purchase or improvements of farm real estate. The absence of effective long-term credit programs was a major gap when the Federal Land Bank System was developed in 1916. A major gap would likely be the case again today in the absence of the Farm Credit System, especially if the public purpose continues to favor a largely pluralistic, smaller-scale structure for the agricultural sector. While farm real estate lending by commercial banks has increased substantially in recent years, it is unlikely that the banking system could completely replace the long-term credit provided by the FCS. A well-developed secondary market for selling farm real estate loans or pooled shares in loans would then be needed (a successful version of Farmer Mac), similar to the case in residential housing, but even then government-sponsored enterprises (Fannie Mae and Freddie Mac) have been needed to make the secondary market work.

The Farm Credit System has also provided an important source of non–real estate credit for farmers that enhances market competition, especially in rural financial markets where commercial banks are less involved in agricultural finance. Reliability and risk-bearing ability are preferred attributes of a local lender. While the FCS needed financial assistance in the 1980s, that case was the first in a fifty-year period. Moreover, the resulting institutional restructuring, strengthening of FCA, upgrading of the system's risk protection mechanisms, and more conservative lending terms were major responses to those troubled times. Finally, the banks for cooperatives have played a major role in the growth and performance of agricultural cooperatives in the United States, including the financing of a modest yet significant amount of international lending.

If further privatization of the FCS is needed, through the removal of agency status, then a broadening of lending, service, and funding authority may be appropriate to consider. Such broadening could enhance risk-carrying capacity, improve services available to borrowers, and fill market gaps in which credit availability to rural communities and business is limited.

As restructuring and consolidations within the FCS continue, the need for the farm credit bank component of the system may ultimately be called into question (Barry, Brake, and Banner 1993). Fewer banks (for example, one, two, or three) and larger lending associations may shift the role of the banks to providing services to the associations and reduce the bank's role in the FCS intermediation process. Lending associations may eventually supplant the banks and remove one of the tiers from the FCS system.

Further restructuring and consolidations may also reduce the importance of the cooperative organization of the FCS. Shifting to outside ownership of stock traded on the public stock exchanges (similar to several other government-sponsored enterprises) would broaden the ownership base, further strengthen risk-bearing capacity, and provide a different perspective and degree of financial discipline among the boards of directors of the FCS institutions. At the same time, however, it would dilute the familiarity with local agricultural conditions that now characterizes the boards of most smaller lending associations.

8

The Federal Agricultural Mortgage Corporation

The Federal Agricultural Mortgage Corporation (Farmer Mac) was created by the Agricultural Credit Act of 1987 to oversee the development of a secondary market for farm real estate loans. Supervised and regulated by the Office of Secondary Market Oversight in the Farm Credit Administration, Farmer Mac operates as an independent entity within the Farm Credit System.

Purpose

The purposes of the secondary market, indicated in the 1987 act, are: (1) to increase the availability of long-term credit to farmers and ranchers at stable interest rates; (2) to provide greater liquidity and lending capacity in extending credit to farmers; (3) to facilitate capital market investments in providing long-term agricultural lending, including funds at fixed rates of interest; and (4) to improve availability of credit for rural housing.

The basic idea of the secondary market is to achieve a separation of loan origination, servicing, funding, and risk bearing, so that the farm mortgage market will function more efficiently. Originally, Farmer Mac was to supervise the purchases by poolers of eligible farm mortgages originated and perhaps serviced by a primary lender. The poolers (organized by life insurance companies, commercial banks, FCS institutions, or others) aggregate the loans into portfolios and then sell pooled participation securities to investors based on a pass-through of principal and interest payments by borrowers or based on securities sold to investors backed by the loan pools. In turn, Farmer Mac guarantees these securities to ensure their safety for financial market investors.

Eligible loans must meet Farmer Mac's standards and be adequately documented. Only farm real estate loans or rural housing loans are eligible. Loan size was initially limited to a maximum of $2.5 million per loan, although this limit is adjusted over time for inflation. Eligible loans must also meet Farmer Mac underwriting standards, which are based on the criteria recommended by the Farm Financial Standards Task Force. To illustrate, the loan-to-land value ratio may not exceed .75; the debt-to-asset ratio may not exceed .50; the borrower's anticipated cash flows must be sufficient to repay the loan; and other ratio and nonratio standards apply.

Several safety mechanisms are contained in the Farmer Mac system to deal with loan delinquencies and losses and thus protect investors in Farmer Mac securities. These mechanisms include: a cash reserve or subordinated participation interest held by originators or poolers equal to at least 10 percent of the loan; a Farmer Mac reserve funded through fees charged to participating financial institutions; Farmer Mac's own equity capital; and a line of credit up to $1 billion with the U.S. Treasury.

Growth Problems

While Farmer Mac was created by the 1987 act, the first pooling operation did not begin until 1991. By 1994, only eight poolers had been certified, and little pooling activity had occurred. The total volume of credit under the Farmer Mac program is in the $800–$900 million range. The slow development has been attributed to several factors: weak loan demand, strong liquidity of agricultural banks, stringent capitalization requirements, uncertain loan volume, questionable interest rate competitiveness, and slow acceptance by investors of the unique and complex features of real estate mortgages in agriculture (U.S. General Accounting Office, September 1991; Freshwater 1992). These factors are reviewed below.

The proposal to establish a secondary market for farm real estate loans began in the early 1970s when loan demand by farmers was strong and the secondary markets for residential housing loans were viewed as successful operations. Commercial banks at that time were seeking an effective way to fund long-term loans to agricultural borrowers. The financial stresses of the 1980s, however, caused a significant downturn in farm loan demand and tightening of credit standards that continued into the 1990s. The reduction in loan demand together with moderate growth in deposits created relatively high liquidity for agricultural banks and thus a diminished incentive to sell loans into a secondary market.

The incentive for loan sales was also reduced when regulatory agencies required banks and FCS institutions to meet equity capitalization requirements (that is, maintain equity reserves) on the full amount of the sold loan rather than on the 10 percent portion of the loan held by the originating lender. The regulators believe that, despite selling 90 percent of the loan, the originator was still carrying all the loan's credit risk.

Potential poolers have also been reluctant to participate in the Farmer Mac program because they are concerned that the volume of loans would not be sufficient to justify their capital commitments to the program. Part of that concern involved the potential lack of competitiveness of interest rates and other lending terms for borrowers. The accumulation of reserves needed to provide the safety mechanisms cited above comes mostly from interest rates paid by borrowers; thus, safety for securities investors may be coming at the cost of less-than-competitive rates for borrowers.

Expanded Authorizations

Beginning in 1990, Farmer Mac received two new authorities intended to stimulate and expand the development of the secondary market. The first new authority allowed Farmer Mac to serve as the pooler for secondary sales of loans guaranteed by the Farmers Home Administration. This activity is called Farmer Mac II, while the original program is Farmer Mac I.

Under Farmer Mac II, originators can sell the FmHA-guaranteed portion of operating or farm ownership loans with maturities of at least one year. The loans are sold for cash or swapped for a marketable security. The sale terms are structured to provide lenders with the current market yield on the funds invested in the loan as well as fee income for continuing to service the loan. The loans may have fixed or variable interest rates; a variable rate loan is adjusted periodically in response to changes in a specified cost-of-funds index. Similar to Farmer Mac I, Farmer Mac II offers lenders the opportunity to increase liquidity and lending capacity and reduce interest rate risks.

The volume of FmHA-guaranteed loans sold through Farmer Mac II totaled $47.6 million in 1994, up from $39.5 million in 1993 (USDA 1994). Despite the growth, the 1994 figure is less than 5 percent of FmHA's fiscal 1994 guaranteed-loan volume eligible for Farmer Mac II sale. Either the program has significant growth opportunity, or it is not viewed as an attractive alternative by most lenders using the loan guarantee program.

The second change in late 1991 authorized Farmer Mac to fund loan pools by issuing its own unsecured debt securities, having agency status similar to that of other government-sponsored enterprises. Under this program, called the linked portfolio strategy, the proceeds of the security sales are used to buy and hold other securities backed by qualifying pools of farm mortgage loans issued by certified Farmer Mac poolers and guaranteed by Farmer Mac. Interest rate risks are managed by closely matching terms and rates on Farmer Mac debt obligations with the terms and rates on the mortgage-backed securities purchased from poolers and by other methods of handling prepayment risk.

The linked portfolio strategy arrangements allow Farmer Mac to function more like Fannie Mae and Freddie Mac. The nature of Farmer Mac's role differs substantially between the Farmer Mac I and the linked portfolio strategy programs. Under Farmer Mac I, Farmer Mac is not directly involved in the intermediation process; however, under the linked portfolio strategy, Farmer Mac does directly take a position in intermediation by issuing its own debt securities.

The linked portfolio strategy attracted renewed pooler interest. Two new poolers began to purchase complete farm loans from originators in 1993, under the terms of the linked portfolio program, with a variety of maturities and pricing methods for borrowers. Little securitization had occurred by the end of 1994, however.

Finally, during 1994 Farmer Mac undertook several initiatives to spur poolers to securitize loans (U.S. Department of Agriculture 1994). Farmer Mac now requires poolers to submit loan pools for guarantee or risk losing certification. Farmer Mac also entered into an alliance with the Western Farm Credit Bank of Sacramento in which the Western FCB will establish and operate a nationwide pooling program for agricultural mortgages open to all Farmer Mac stockholders. The success of these initiatives also remains to be seen. In contrast, most sales of FmHA-guaranteed loans now occur under Farmer Mac's linked portfolio program.

Farmer Mac activities are also receiving a boost through the formation of a network of minipoolers. These minipoolers provide a bridge between loan originators (primarily smaller banks) and larger poolers. They also play a major role in applying Farmer Mac underwriting standards to evaluate eligibility of loans for pooling. The degree of success of the minipooler concept remains to be seen.

Outlook

The slow development of Farmer Mac has raised serious concerns about the program's long-term viability, even with the recent development of the Farmer Mac II, linked portfolio strategy, and minipooler components. Weak loan demand, strong liquidity in agricultural banks, stringent capitalization requirements, and concerns about competitiveness are plausible and significant issues. The ultimate demand for the program, in an environment of stronger growth in loan demand, remains unclear.

Other questions, issues, and perspectives about the future of Farmer Mac are also important to consider. It is interesting, for example, that commercial banks were able to increase their volume of farm real estate loans substantially—from $7.6 billion in 1982 to $19.5 billion in 1993 (while aggregate farm debt was declining)—without the assistance of the Farmer Mac program. Some of the growth came at the time of FCS problems and when fund availability from banks was strong. But the loan volume growth still is substantial.

Moreover, the creation of Farmer Mac in 1987 occurred in legislation directed primarily toward the financial recovery of the Farm Credit System. To some extent, the major competitors (that is, life insurance companies and commercial banks) of the FCS used the circumstances of those times and the political process to gain a new financial innovation (Farmer Mac) as a concession from the FCS to provide federal financial assistance to aid in its own recovery. Thus, Farmer Mac was created, in part, as a quid pro quo rather than as a response to a strong need-driven demand.

Perhaps the most fundamental question is whether two government-sponsored enterprises, the FCS and Farmer Mac, are needed to make the farm real estate loan market work effectively. The dominant activity of, and contribution by, the Farm Credit System is farm real estate lending. FCS real estate lending usually occurs at competitive interest rates that reflect the funding cost benefits of agency status of FCS securities and an efficient, cost-effective loan delivery system. Moreover, as noted earlier, the FCS now has an extensive set of protection mechanisms to offset the risks of specializing in agricultural loans. If changes in the agency status of FCS securities resulted in a greater privatization of the system, then the secondary market role of Farmer Mac might take on greater significance. Farmer Mac might also receive a boost if the FCS institutions take greater interest in the secondary market as a strategic marketing tool.

The development of the secondary market for farm mortgage loans was a major financial innovation for agricultural finance. In addition, the programmatic changes of Farmer Mac since 1990, the continued restructuring of the Farm Credit System, further regulatory changes in the geographic scope of commercial banking, and efforts to downsize FmHA programs all indicate that the financial markets for the agricultural sector are experiencing significant transition. It is too early to abandon Farmer Mac as only an interesting, yet failed experiment. We need to see how its potential role may change as the financial markets for agriculture continue to evolve.

9
Commercial Banking

Discussion earlier in this volume established the dominant position of commercial banks in financing agriculture, including the strong yet diminishing role played by small agricultural banks. While banks are highly regulated financial institutions, few elements of banking statutes or regulations are directed specifically toward the financing of agriculture. Nonetheless, future structural changes in banking could lead to reductions in the number and extent of involvement in agricultural lending by small banks and to increased demands for public credit programs (Barry 1981).

The nearly 4,000 agricultural banks in the United States out of about 11,000 banks in total at the beginning of 1994 have successfully weathered the effects of the 1980s financial stresses in agriculture. Bank failures over the 1981–1993 period totaled 360 for agricultural banks and 1,086 for nonagricultural banks. Rates of return on equity of these small banks have also been restored to the 12–14 percent range in recent years after several years of single-digit rates in the 1980s. These banks have also responded successfully to the greater competition in financial markets attributed to interest rate deregulation and to substantial liberalization of the products and services banks may offer. They have also survived, so far, the geographic liberalization of banking mostly involving reciprocal holding company legislation among states and lifting of branch-banking restrictions in several formerly unit-banking states. Most of the past holding company expansions have involved money center or regional banks, with less involvement by small rural banks.

Another significant boost to geographic restructuring of banks will come from the 1994 legislation allowing interstate banking on a national basis, unless individual states choose not to participate. It is widely anticipated that the number of individual banks in the United States will decline substantially over the next five to ten years; however, the number of bank offices may remain about the same. This change will mean fewer agricultural banks, although it is also well recognized that agricultural lending will remain an important niche market for well-capitalized, highly competitive, and strongly managed small banks that develop effective relationships for using the services provided by larger bank and nonbank financial systems.

A key question for public credit programs is whether continued consolidation in banking will significantly affect the cost and availability of credit and other financial services for agricultural borrowers, thus expanding the demand for public credit. In general, lower market shares of farm debt are typically held by banks in the Southeast, delta states, and the Northeast, partly because of the larger, more metropolitan banking systems in these regions of the United States (Barry et al. 1995). Wilson and Barkley (1988) also found that bank structure is a significant variable in explaining the differences in changing market shares across states. Their results indicate the agricultural lending levels for commercial banks adjusted downward to a lesser extent in unit- and limited-branching states than in statewide branching states between 1969 and 1982.

Statewide branching provides banks with greater diversity in their loan portfolios and less reliance on agricultural lending.

Similarly, Belongia and Gilbert (1988) observe that affiliation with multibank holding companies significantly reduces the ratio of agricultural loans to total loans for the rural subsidiaries of large bank holding companies. The subsidiaries of large bank holding companies have greater opportunities to diversify risk by lending to businesses in a variety of industries, thus reducing the supply of agricultural credit through commercial banks. An offsetting factor was observed by Laderman, Schmidt, and Zimmerman (1991), who found that, when statewide branching is permitted, rural banks hold higher nonagricultural loan portfolio shares and urban banks hold higher agricultural loan portfolio shares. Their results offer support for the hypothesis that branching enhances diversification and a broader basis of support than can be obtained from a study of rural banks alone.

If continued geographic liberalization of banking tends to reduce the supply of agricultural credit, the burden will probably fall on smaller, higher-risk farmers, who will then seek public credit. The extent of this burden may in turn be determined by the ability of commercial banks to distinguish among their various classes of borrowers (that is, resolve asymmetric information problems). Improvements in credit evaluation procedures, including the use of credit-scoring models, allow better distinctions to be drawn, as long as borrowers can furnish the necessary financial information. Doing so seems more problematic for smaller borrowers.

10
Concluding Comments

Both agriculture and its financial markets are continuing to experience substantial transition. A tri-modal structure is emerging in agriculture characterized by the coexistence of large industrialized units, commercial-scale family operations, and small, part-time, or limited-resource farmers. Commercial lenders, including the Farm Credit System, are meeting the financing needs of the industrialized units. These units neither need nor use subsidized credit programs, except perhaps for some younger, inexperienced contract growers in poultry and related types of production. Commercial-scale family operations and small farms are financed by a variety of sources, including public credit programs. For these producers, the availability of public credit seems to play a more crucial role. The traditional mid-sized family farms, which have been a major focus of FCS lending, are diminishing in number and relative importance.

In light of the changing characteristics of the agricultural sector, it is appropriate to monitor parallel types of changes in the need for and availability of financial capital. Agriculture has an extensive set of credit sources: the banking system and other commercial lenders, two government-sponsored enterprises, and a government-owned lender of last resort. These public credit programs have responded to each of the major rationales for public credit. The FCS, which filled a major real estate financing gap, also provides a reliable, specialized, competitive source of operating and capital credit for farmers, ranchers, and their cooperatives. The specialization has helped overcome the information problems and relatively high transactions costs of agricultural lending. Without the FCS, the void in farm real estate lending would be substantial. The secondary market operations of Farmer Mac would become more important. Farmer Mac was created to improve the workings of the farm real estate credit market, although the need for two separate government-sponsored enterprises—one for direct lending and another for secondary transactions—is an interesting issue.

FmHA has served the public purposes of facilitating resource adjustments in agriculture, providing liquidity in times of adversity, and assisting many agricultural borrowers in meeting the creditworthiness requirements of commercial lenders. In the process, FmHA has provided substantial subsidies, although subsidization has been reduced by the shift to guaranteed loans. FmHA, however, is especially vulnerable to the political economy of credit programs that easily lead to institutional overload. The fuzziness and varying nature of its mission, especially in the 1980s, have led to weak loan quality (even for a lender of last resort), slow graduation of borrowers to commercial credit, management and control problems, moral hazards by borrowers and lenders, and a major welfare role. Some observers (for example, Herr 1994) even question the continuing need for these programs. How FmHA programs will fare in the proposed USDA restructuring is unclear.

Key questions remain about public preferences, if any, toward the structural features of the agricultural sector. The FCS and FmHA play a major role in preserving the traditional pluralistic, small-scale organizational and ownership structure of the sector. But the forces of industrialization are working against this traditional view. If these forces continue to work their course, then the traditional scope, missions, and operations of the FCS and FmHA are subject to change as well. Changes will not likely occur overnight, but the journey may have begun.

References

Arthur, L., C. Carter, and F. Abizadeh. "Arbitrage Pricing, Capital Asset Pricing, and Agricultural Assets." *American Journal of Agricultural Economics* 70(1988):359–65.

Barkema, A., and M. Cook. "The Changing U.S. Pork Industry: A Dilemma for Public Policy." *Economic Review, Federal Reserve Bank of Kansas City*, vol. 78, no. 2 (1993b):49–66.

———. "The Industrialization of the U.S. Food System." *Food and Agricultural Marketing Issues for the 21st Century.* Food and Agricultural Marketing Consortium, FAMC 91-1, Texas A&M University, 1993a.

Barkema, A., M. Drabenstott, and K. Welch. "The Quiet Revolution in the U.S. Food Market." *Economic Review, Federal Reserve Bank of Kansas City*, vol. 76, no. 3 (1991):25–41.

Barry, P. "The Farmers Home Administration: Current Issues and Policy Directions." *Looking Ahead.* National Planning Association, Washington, D.C. 8(September 1985):4–12.

———. "Capital Asset Pricing and Farm Real Estate." *American Journal of Agricultural Economics* 62(1980):549–53.

————. *Impacts of Financial Stress and Regulatory Forces on Financial Markets for Agriculture.* National Planning Association, Washington, D.C., 1984.

————. "Impacts of Regulatory Change on Financial Markets for Agriculture." *American Journal of Agricultural Economics* 63(1981):905–12.

————. "Needed Changes in the Farmers Home Administration Lending Programs." *American Journal of Agricultural Economics* 65(1985):341–44.

Barry, P., and M. Boehlje. "Farm Financial Policy." D. Hughes, S. Gabriel, P. Barry, and M. Boehlje. *Financing the Agricultural Sector.* Boulder, Colo.: Westview Press, 1986.

Barry, P., and W. Lee. "Financial Stress in Agriculture: Implications for Agricultural Lenders." *American Journal of Agricultural Economics* 65(1983):945–52.

Barry, P., and L. Robison. "Economic versus Accounting Rates of Return on Farm Land." *Land Economics* 62(1986):388–401.

Barry, P. J., J. R. Brake, and D. K. Banner. "Agency Relationships in the Farm Credit System: The Role of the Farm Credit Banks." *Agribusiness* 9(1993):233–45.

Barry, P. J., P. N. Ellinger, J. Hopkin, and C. Baker. *Financial Management in Agriculture,* 5th ed., Danville, Ill.: Interstate Publishers, 1995.

Belongia, M., and A. Gilbert. "The Effects of Affiliation with Large Bank Holding Companies on Commercial Bank Lending to Agriculture." *American Journal of Agricultural Economics* 70(1988):69–78.

Bjornson, Bruce. "Asset Pricing Theory and the Predictable Variation in Agricultural Asset Returns." *American Journal of Agricultural Economics* 76(1994):454–64.

Bosworth, B., A. Carron, and E. Rhyne. *The Economics of Federal Credit Programs.* Washington, D.C.: Brookings Institution, 1987.

Budget of the United States Government. *Analytical Perspectives FY1995.* U.S. Government Printing Office. Washington, D.C., 1994.

Christomo, M. F., and A. Featherstone. "A Portfolio Analysis of Returns to Farm Equity and Assets." *North Central Journal of Agricultural Economics* 12(January 1990):9–22.

Council on Food, Agricultural, and Resource Economics. *Industrialization of U.S. Agriculture: Policy, Research, and Education Needs.* Washington, D.C., April 1994.

Ellinger, P., and P. Barry. "The Effects of Tenure Position on Farm Profitability and Solvency." *Agricultural Finance Review* 47(1987):106–18.

Ellinger, P., N. Splett, and P. Barry. "Consistency of Credit Evaluations at Agricultural Banks." *Agribusiness: An International Journal* 8(1992):517–36.

Freshwater, D. "Farmer Mac: An Idea Whose Time May Never Come." Staff paper no. 316, Department of Agricultural Economics, University of Kentucky, Lexington, January 1992.

Gale, D., and M. Hellwig. "Incentive Compatible Debt Contracts: The One Period Problem." *Review of Economic Studies* 52(1985):647–63.

Gale, W. G. "Economic Effects of Federal Credit Programs." *American Economic Review* 81(1991):133–52.

———. "Federal Lending and the Market for Credit." *Journal of Public Economics* 42(1990a):177–93.

———. "Collateral, Rationing, Government Intervention in Credit Markets." *Asymmetric Information, Corporate Finance, and Investment,* edited by R. G. Hubbard. Chicago: University of Chicago Press, 1990b.

———. "The Allocational and Welfare Effects of Federal Credit Programs." Ph.D. diss., Stanford University, 1987.

Gustafson, C., and P. Barry. "Structural Implications of

Agricultural Finance." In *Size, Structure and the Changing Face of American Agriculture,* edited by A. Hallam. Boulder, Colo.: Westview Press, 1993.

Herr, W. *Toward an Analysis of the Farmers Home Administration's Direct and Guaranteed Farm Loan Programs.* Economic Research Service, U.S. Department of Agriculture, April 1991.

————. "Are Farmers Home Administration's Farm Loan Programs Redundant?" *Agricultural Finance Review* 54(1994):1–14.

Herr, W., and E. LaDue. "The Farmers Home Administration's Changing Role and Mission." *Agricultural Finance Review* 41(1981):58–72.

Irwin, S., L. Forster, and B. Sherrick. "Returns to Farm Real Estate Revisited." *American Journal of Agricultural Economics* 70(1988):580–87.

Jaffee, D., and T. Russell. "Imperfect Information, Uncertainty, and Credit Rationing." *Quarterly Journal of Economics* 90(1976):651–66.

Koenig, S., and J. Stam. *Life Insurance Company Farm Lending during the 1980s: Evolution or Revolution.* Unpublished paper, Economic Research Service, U.S. Department of Agriculture, 1992.

Laderman, E., R. Schmidt, and G. Zimmerman. "Location, Branching and Bank Portfolio Diversification: The Case of Agricultural Lending." *Economic Review.* Federal Reserve Bank of San Francisco (Winter 1991):24–37.

LaDue, E. "Moral Hazard in Federal Farm Lending." *American Journal of Agricultural Economics* 72(1990):774–79.

Lee, J., and S. Gabriel. "Public Policy toward Agricultural Credit." *Future Sources of Funds for Agricultural Banks.* Federal Reserve Bank of Kansas City, Kansas City, 1980.

Lins, D., and P. Barry. "Agency Status of the Farm Credit System." *American Journal of Agricultural Economics*

66(1984):601–6.

Mankiw, G. "The Allocation of Credit and Financial Collapse." *Quarterly Journal of Economics* 101(1986):455–70.

Meekhof, R. *Federal Credit Programs for Agriculture,* Agricultural Information Bulletin 483. Economic Research Service, U.S. Department of Agriculture, November 1984.

Melichar, E. "Capital Gains vs. Current Income in the Farming Sector. *American Journal of Agricultural Economics* 61(1979):1082–92.

Office of Management and Budget. "Special Analysis F" in Special Analyses: Budget of the United States Government, selected years.

Patrick, G., P. Wilson, P. Barry, W. Boggess, and D. Young. "Risk Perceptions and Management Responses: Producer-generated Hypotheses for Risk Modeling." *Southern Journal of Agricultural Economics* 17(December 1985):231–38.

Penner, R. "Comments on Credit Rationing and Government Loan Programs: A Welfare Analysis." *AREUEA Journal* (American Real Estate and Urban Economics Association) 17(1989):194–96.

Rhodes, V. J., and G. Grimes. "U.S. Contract Production of Hogs: A 1992 Survey." Report 92-2. Department of Agricultural Economics, University of Missouri, 1992.

Smith, B., and M. Stutzer. "Credit Rationing and Government Loan Programs: A Welfare Analysis." *AREUEA Journal* 17(1989):177–93.

Stiglitz, J., and A. Weiss. "Credit Rationing in Markets with Imperfect Information. *American Economic Review* 71(1981):393–410.

U.S. Department of Agriculture. *Agricultural Income and Finance.* Situation and Outlook Report. Economic Research Service, AIS-52, February 1994, and AIS-56, February 1995, Washington, D.C.

U.S. General Accounting Office. *Farmers Home Adminis-*

tration: Billions of Dollars in Farm Loans Are at Risk. GAO/ RCED-92-86, Washington, D.C., April 1992.

————. *ADP Modernization: Half-Billion Dollar FmHA Effort Lacks Adequate Planning and Oversight.* GAO/ IMTEC-92-92, Washington, D.C., October 1991.

————. *Federal Agricultural Mortgage Corporation: Secondary Market Development Slow and Future Uncertain.* GAO/RCED-91-181, Washington, D.C, September 1991.

Wilson, P., and D. Barkley. "Commercial Banks Market Shares: Structural Factors Influencing Their Decline in the Agricultural Sector." *Agricultural Finance Review* 48(1988):49–59.

Young, R., and P. Barry. "Holding Financial Assets as a Risk Response." *North Central Journal of Agricultural Economics* 9(1987):77–84.

Index

About the Author

PETER J. BARRY is professor of agricultural economics at the University of Illinois and has served as president of the American Agricultural Economics Association. He is also director of the Center for Farm and Rural Business Finance. His research activities include studies of credit evaluation in agriculture, farm financial management, risk management, performance of financial markets for agriculture, asset-liability management, and loan pricing by financial institutions, asset valuation, and investment analysis. Mr. Barry was editor of the *American Journal of Agricultural Economics* and the *Western Journal of Agricultural Economics,* and he was a faculty member at Texas A&M University and at the University of Guelph. He holds honors from the International Confederation of Agricultural Credit, Istanbul, Turkey, and the College of Agriculture, University of Illinois.

A NOTE ON THE BOOK

This book was edited by Dana Lane
of the publications staff
of the American Enterprise Institute.
The figures were drawn by Hordur Karlsson.
The index was prepared by Robert Elwood.
The text was set in Palatino, a typeface
designed by the twentieth-century Swiss designer
Hermann Zapf. Lisa Roman of the AEI Press
set the type, and Braun-Brumfield, Inc.,
of Ann Arbor, Michigan,
printed and bound the book,
using permanent acid-free paper.

The AEI Press is the publisher for the American Enterprise Institute for Public Policy Research, 1150 Seventeenth Street, N.W., Washington, D.C. 20036; *Christopher DeMuth*, publisher; *Dana Lane*, director; *Ann Petty*, editor; *Leigh Tripoli*, editor; *Cheryl Weissman*, editor; *Lisa Roman*, editorial assistant (rights and permissions).

www.ingramcontent.com/pod-product-compliance
Lightning Source LLC
Jackson TN
JSHW011940131224
75386JS00041B/1477